THE COMPANY WE KEEP

THE COMPANY WE KEEP
Reinventing Small Business for People, Community, and Place

JOHN ABRAMS

FOREWORD BY WILLIAM GREIDER

CHELSEA GREEN PUBLISHING COMPANY
WHITE RIVER JUNCTION, VERMONT

Managing Editor: Collette Leonard
Project Editor: Marcy Brant
Developmental Editor: Woody Tasch
Copy Editor: Nancy Ringer
Proofreader: Laura Jorstad
Indexer: Beth Nauman-Montana, Salmon Bay Indexing
Designer: Peter Holm, Sterling Hill Productions
Design Assistant: Daria Hoak, Sterling Hill Productions

Printed in the United States on recycled paper
First printing, April 2005
10 9 8 7 6 5 4 3 2 1

Library of Congress Cataloging-in-Publication Data
Abrams, John.
The company we keep : reinventing small business for people, community, and
place / John Abrams ; foreword by William Greider.
    p. cm.
Includes bibliographical references and index.
ISBN 1-931498-73-3
1. South Mountain Company—History. 2. Construction
industry—Massachusetts—Martha's Vineyard—Case studies. 3. House
construction—Massachusetts—Martha's Vineyard—Case studies. 4. House
construction—Environmental aspects. 5. Social responsibility of business.
6. Industrial management—Environmental aspects. I. Greider, William. II.
Title.
HD9715.U54S682 2005
338.7'6908'0974494--dc22

                              2005000484

Chelsea Green Publishing Company
Post Office Box 428
White River Junction, VT 05001
(800) 639-4099
www.chelseagreen.com

To my wife Chris, the bravest person I know.

CONTENTS

FOREWORD

What you have in your hands is the memoir of one small residential design and building company, written by the cofounder who has been the devoted steward of the enterprise across some thirty years. Improbable as this may sound, you will find his story irresistibly charming. It is also, perhaps, challenging to the usual ways of thinking about businesses (and building contractors). But, above all, it is intellectually provocative, because this book takes the reader from the intimate, pedestrian details of company life to much grander propositions about society and the elements of a sound economy.

There is a seductive quality in this. You start with an engagingly modest story about one small place that tried to do things differently. Then you find yourself thinking about the nature of honest work and cooperation, about the deeper meanings of quality and enduring value, about designing and creating self-accountability among the people who share in the life of a firm.

John Abrams is one of those rare souls who set out to find his own way in life, without much of a plan but with strong, distinctive personal and social values. He sounds a trifle naive at first because he seems to think that one can adhere faithfully to such convictions while also pursuing commerce. Yet he has succeeded at this (he would probably say he is still working on it). But the story is not really about him; it's about South Mountain Company and how it found its groove and prospered. The author tells enough about himself that you can locate his unselfconscious humility and optimism as authentic. What he really wants you to understand is the living, evolving organism of this company. The message is about how craftsmanship and dignity can be restored and sustained in a world of industrial complexity and the mass-market degradation of cherished values.

South Mountain starts with a handful of people and expands gradually.

Its coworkers become co-owners and work their way through the usual obstacles and temptations of business life. Little by little, however, they find ways—actually the company's precious cornerstones—to sustain a collegial sense of self-criticism. Managed inclusiveness and mutual respect, one might call it. Because Abrams is relentlessly modest (and compassionate) in his storytelling, many of the false starts or wrong ideas are attributed to him and challenged and corrected by others. The business approaches and undertakings are repeatedly held up to the shared values and tested. The process is not without setbacks and compromises. But, even discounting for his modesty, Abrams convinces one that this culture does exist as everyday reality at South Mountain, and that it works.

The story is extraordinary, but what can we learn from it? After all, the company is small and operates in an exotic marketplace—the island of Martha's Vineyard—where the culture and affluence are unusual, and people are disposed to support high principles in home design and building. I began reading the memoir with that thought foremost in my mind. When I finished, I realized, no, that's mistaken. Their experiences are in microcosm, but the ideas and understandings formed among the human beings seem universal. With goodwill and serious values, they may be portable to other places, including larger enterprises, where the people also want to do things differently.

William Greider

THE COMPANY WE KEEP

A welcoming entrance capped with driftwood. (Photo by Brian Vanden Brink.)

1

CORNERSTONES

South Mountain Company, the business I cofounded in 1975 with my friend Mitchell Posin on the island of Martha's Vineyard, makes houses. Big houses, small houses, and neighborhoods, new from scratch and remade from old. Affordable houses and distinctly *unaffordable* houses. Sweet successes and partial failures. We do all the parts—we plan, develop, design, build, finish, and furnish. In the doing we form lasting relationships with our clients and their land. We maintain the buildings after completion, learn from how they work and age, and alter and add when the time comes. Over the past thirty years we have designed and built well over one hundred houses, renovations, and small neighborhoods, and we have become, individually and collectively, active participants in the life of our island community and the planning for its future. It has been a remarkable journey of discovery.

Along the way, as we have become a part of this place, we have come to sense that we are only at the beginning, that our endeavors—and our company—may have, or can aspire to have, some of the enduring qualities of the buildings we fashion. We have made a series of commitments to the future of the people within the company, to those who will come after us, and to the island where we all live and work. We are building a community of enterprise that we hope will have the strength to nurture stable relationships, the flexibility to shift with changing times, and the ability to sustain itself for generations.

I am finding out that people mostly don't think of companies this way.

Companies are entities that people start, capitalize, run, work for, buy and obtain services from, sell, and disband. But South Mountain has become, for us, as much a community as a company. We not only build houses, we build connections and bonds between people, between people and land, and between commerce and place. We are organized around the idea of maintaining and perpetuating an ongoing business community. We *think* we are crafting a company to keep.

If we are lucky in life, work becomes an expression of who we are and one of our most important anchors of meaning. I was lucky enough to discover this early and quite by accident. The story of my coming of age has now become a familiar one—a '60s kid, brash and audacious, who set out to make a new world free of greed and conformity, aligned with nature, grounded in justice for all.

When the 1960s ended I had just turned twenty. I renounced, for a time, modern consumer culture and technology in favor of the skills and tools of rural self-sufficiency. I, like many others, was more at home in America's agrarian past than in mainstream society's present. The journey was not without contradictions: we went deftly from the suburbs toward Walden like hopping across a creek, thinking we could cross back if necessary. But this inconsistency seemed less important than the value of our dogged attempts to imagine new ways that made sense to us.

It was a passage aligned with the sweeping political and social upheaval of the times. We were contending with war in Vietnam and war in the streets. We had experienced the Cuban missile crisis as teenagers and huddled beneath our desks as schoolchildren.

My wife has a favorite comic strip in which a son says to his father, "Now wait a minute, let me get this straight. You're saying that in the school you went to they used to tell you in the event of nuclear attack you should get under your desk?"

The father says, "That's right."

In the next panel the son asks, "Would that be the same school where you learned everything else you know?"

We had similar questions. The search for answers led us to unexpected places from which the return seemed increasingly unlikely. A new culture was so much in the air that our identification with it was almost like slipping into a pair of comfortable shoes. I don't know the word that would best describe what we were then—*network*, *tribe*, or *clan*—but wherever we traveled we felt a shared collective consciousness that was a product of our profound alienation from the values we saw reflected in society combined with exquisite anticipation of the invention of a new way of living, a new kind of home.

Crash Course

Immersed in our new way of life, distant from our suburban roots, we made shelter by fashioning crude tepees and fixing leaky roofs in abandoned farmhouses. I began to have an urge to understand buildings. I marveled at ancient barns, walked on their hewn timbers, studied how they were joined, and sometimes dismantled them piece by piece. Old Capes, Colonials, Victorians, and bungalows became at once church and school: places where I could gaze with pleasure through tiny panes of handblown glass, run my hands over undulating horsehair plaster walls, poke around in attics finding the stuff that lay behind lime-mortared chimneys, and learn secrets from those who made these buildings long ago.

Root cellars, woodsheds, corncribs, blacksmith shops, junkyards, stone walls, rusted farm implements, and especially the old-timers who loved a wide-eyed audience—all contributed to our new storehouse of arcane knowledge. We rummaged through bins at farm sales, searching for old tools like broadax and adze for hewing, froe for splitting, peavey and cant hook for wrestling logs around, and beetle for persuading uncooperative timbers to fall into place. Grateful for the remnants of another time, we hauled them home to test them out and learned to use them to live and work in the rural outbacks of Vermont, northern California, Oregon, and British Columbia. Our lives then were devoted to the basics: growing and preparing food, raising animals, finding water, making and mending clothing, creating shelter, fabricating toys for our young ones, entertaining ourselves. There was no separation between work and play. Building and design gradually became particular passions of mine, and here too there was no distinction—each of the two was an integral part of a whole. Our newfound ethic of self-sufficiency dictated that we learn by the seat of our pants, far away from conventional classrooms, no matter how difficult the challenges were and how unattainable the ideal of self-sufficiency.

I was lucky to have begun my work with a partnership that was far more than a business relationship. Mitchell was my closest friend. We not only worked side by side but also traveled and lived together during those years. He came from Brooklyn, and his working-class background, which included repairing furniture with his father and laboring during his teens at a riding stable (yes, there were still horses in Brooklyn), was the perfect complement to my academic upbringing. He had the ability to think like a craftsman and work with his hands, and I brought the ability to research, scheme, and absorb knowledge by reading. We learned from each other.

The first house we built was in Vermont in 1972. We heard through the grapevine that Harry Saxman, who lived up the road, was planning to build a house. We went to see him.

"We hear you're planning to build a house," we said, "and we're wondering if you've got someone to build it."

"Can you guys build a house?" he asked.

"Well, sure, yes, we can."

He hired us. He did the contracting and we did the building. We knew some carpentry and something about buildings, but we didn't have the foggiest notion how to build a house. Each day we worked hard at the job site; each night we spread out on the living room floor of the farmhouse we were caretaking with a stack of carpentry and building books, desperately trying, by flickering kerosene light, to figure out how to do the next day's work. Talk about seat of the pants; this was a crash course. Somehow the house got built.

One Job

In my youth I had few jobs and never stuck long with any of them. As a kid, I had a paper route that I loved. Flinging papers from my bicycle perch and trying to "porch 'em" just right was a delight. I had no summer jobs in high school that I remember. When I was in college I worked one summer in a brick kiln . . . for a few weeks. And in Colorado I had a job at a sawmill . . . for a few more. When I tired of Marlboro College (after one semester), instead of getting a job I opened a hippie store in Brattleboro, Vermont.

We sold artwork and crafts on consignment, drug paraphernalia, and strange clothing, and ours was about the only place in town, at the time, to buy records, both new and used. Later, while I was at Wesleyan University in Connecticut, a friend and I started a firewood business. I worked odd jobs from the communes I lived in: dismantling buildings and working for a surveyor in Vermont, clearing brush in British Columbia, making oak firewood and cedar fence posts in Oregon, doing small carpentry jobs wherever I went.

My first real job came in 1973. I was doing some woodwork for my father-in-law and needed some rough oak planks planed. I took them to a nearby woodworking shop, Krager Custom Woodwork, located in a cavernous space in a complex of old mill buildings near the Hudson River in Garnerville, New York. I fell in love with the place the moment I walked in. Sawdust was everywhere: it hung in the air as if suspended in time, its aroma perfumed the space, and a thin coating softened the oversize, antiquated industrial machinery. Each piece of wood in the scattered piles seemed to clamor to be the next one chosen, anxious to become the next cabinet or coffee table. I didn't want to leave. After the foreman had finished planing, I piled the boards into my truck and went back inside, to the office, and introduced myself to Dan Krager.

"I want to come to work here," I announced.

"We really don't have that much work that we need to hire someone else," he replied.

"I'll work cheap and I'll work hard. I'll do whatever you ask. Just give me a shot."

The way I remember it, I told him that if he didn't give me a job, I'd just hang around and be a pain in the neck, so he might as well. I'm not sure it went quite that way, but he did give me a job.

It was thrilling. I'd wake up excited each morning, hop out of bed, throw on my clothes, and tear over to spend another day belt sanding or pushing wood through a table saw. It was hard, boring work, and I loved every minute of it. There were only four of us in the shop, so I got to witness—and mostly be a part of—each step of every process and project. It was a tremendous learning experience until, only months later, Dan called in the three of us and told us he was going out of business. He just wasn't making it.

It was a sad day, and for me it was the end of what has turned out to be the only real experience I ever had working for a paycheck, for a boss, for more than a few weeks. Meanwhile, I had become friendly with the shop foreman, Kingsley. He and I decided to start doing cabinetry on our own. We bought some equipment at the Krager shop auction, and the first iteration of South Mountain Woodwork began. I called Mitchell, who was in western Massachusetts, and asked him to come down to help build a new shop and work with us. He did, and we were soon off and running. It went well for a year or two, but when the opportunity to build a house on Martha's Vineyard came up, Mitchell and I departed. Kingsley stayed behind and continued the cabinet business.

It turns out, now, that aside from that brief interlude at Krager's, my only real job has been my work at South Mountain Company on the Vineyard. It also turns out that my work has been so completely enmeshed with all the parts of my life that it has never felt different from play. I remain uncertain about the difference.

Stumbling into Business

Mitchell and I did not originally intend to go into business; we stumbled into it by accident while we meant to be learning to practice a craft and explore our passion for building. We didn't mean to be on the island of Martha's Vineyard, either; that, too, happened almost by mistake. We came to the Vineyard in 1975 to make a house for my parents. My mother and father had the impression that we could actually design and build them a house, despite our relative lack of experience. Their generous faith expanded our abilities, but we bit off more than we could chew. We planned to spend six months building and then leave the Vineyard with a pocketful of change, but twelve months later we were still hustling to finish this detailed, timber-framed house with handmade doors, windows, cabinetry, and built-ins. The money had long since run out. While we were struggling through the project, new opportunities came our way.

We had no intention of staying, but in the fall of that year we set up shop in Roger Allen's old barn at Allen Farm. We found ourselves designing and building more houses on the Vineyard, and losing money

The shop at the Allen Farm with Chris's old Valiant and sleeping dog in front. (Photo by Danny Sagan.)

on each one. Perhaps that provided us with moral justification for what we were doing. Our mission was twofold: to keep working and to build the perfect house. We managed to accomplish the first.

A defining moment for me happened during a visit of a friend and mentor who closely followed our work. We were touring our projects. He admired the work. "It's good," he said. "Beautiful work. Making any money?"

"No way," I laughed.

"Nice idea, Abrams," he said. "Novel, anyway. Subsidized housing for the rich."

His blunt assessment triggered the realization that I needed to learn about business, but I wasn't thrilled with the prospect. "You've gotta be kidding," I thought. "I'm still learning the basics of design and can't even make a credible set of working drawings yet. And building? Framing an unequal pitch valley is as mysterious as the darkest jungle. Now this?"

But it was clear to me that our design/build stool would not stand without a sturdy business management leg. So I began to learn, in my own random and ragged way, and it would lead to interesting places, unusual findings, and lucky discoveries.

There was plenty to learn: estimating, ordering, contract writing, scheduling, and insurance. There was bookkeeping and banking to do, cash flow and credit to consider. I traversed back and forth through Bernard Kamaroff's classic *Small-Time Operator*[1] (now in its twenty-fifth edition and still the best—and most irreverent—basic small-business instruction manual). Then there were taxes: income, sales, and excise. Kamaroff has a chapter in his book called "Deep in the Heart of Taxes." I didn't want to visit.

But I came to understand that if our business foundation was level and

Originals in the early '80s. (Photo by Danny Sagan.)

strong, we could build on it. If not, we were doomed to repeat each mistake. The work continued to fall in our laps as people saw the reasonably good buildings we were making, but we had grown tired of subsidizing. So we shuffled paper, created systems, hired a bookkeeper and an accountant, and gave the same respect to our spreadsheets as we did to our chisels. We learned about the utility of profits. The stool gained balance. Greater business understanding brought new freedoms. I learned that business is a craft just like making a staircase. This simple recognition was fuel for the journey.

Some of our best business lessons came from our clients. Eli Sagan, a great supporter, taught us about profit. When we had completed the design of his family's house and were working on the contract he said, "I want to be sure you make a profit on this house, because you're in business and you must, but I'm not willing to pay more than I should. So I'm going to teach you how to make a profit." He did. He went through our estimate line by line with us and asked us the right questions to make sure we had the bases covered. It was lucky for all of us.

Step by step and lesson by lesson our casual, impromptu practice evolved into an organization with a new vantage point. We were used to looking critically at what we designed and how we translated intention to building. Now we could look at ourselves as an entity and examine the kind of *business* we were designing and building. Our principles, which we had invested in the buildings, began to steer the business as well.

One Voice

In 1984 my partner Mitchell went on to other things—farming especially[2]— but the company endured. My colleagues and I have now been making houses on this small island for my entire adult life. From the doing I've learned something about buildings, business, and life in a small community. There are lessons from all three I'd like to forget, and others I wish I could remember, but mostly I've learned that small enterprises, if driven as much by principled practice as by profit, can produce workplace satisfaction, support good lives, and help shape strong communities. This learning has led those of us at South Mountain to a fundamental purpose: to use our business as a tool to help us create these good things. Over time both the community we work in (the island) and the community we have made (the company) have come to matter to us as much as the work we do.

You may think this story I am going to tell is too exotic, too quixotic, to have much meaning for others. Here's my hope: that you will see that South Mountain is, perhaps, a working model for our larger aspirations, and more typical than you might guess. The story tells what I learned with my colleagues and from them, the mistakes we made and our accidental discoveries, and the principles we found to guide us. These all speak to the larger struggle many people are facing today: how to live properly and profitably in concert with their values.

This book is about business, but rather than concentrate on the troubles and injustices of globalization and corporate excess (others are doing that well[3]), I will focus on eight cornerstone principles we have discovered, and how they indicate the promise of small business as a counterforce to the conventional "bigger is better, profits come first, and location is incidental" message. For the sake of good work and good community I will question some cherished assumptions about business, argue for broader and deeper measures of success, and tender alternatives that have worked for us—so far—and may work for others.

Author William Greider is one of those who speaks well about the troubles of our economic system. In *The Soul of Capitalism* he proposes that our economic system is most likely to be transformed, over time, by a variety of small-scale reformers working from within, experimenting with new forms of business in which Americans own their work, their voice, and their self-expression. He believes that this change is already under way. I like to think that my colleagues and I are among those sharing this adventure.

If a lack of believable models is part of what inhibits change, perhaps the story of South Mountain Company—and small, democratic, community-based companies like it—can provide inspiration, or at least information, for those looking for a more desirable approach to business success. Ours is not a cry for help or recognition from a lonely wilderness but rather one optimistic voice in a far-flung choir. There are many others on this journey. Our paths follow those of adventurers who took similar routes in earlier times, and from whose journeys we have learned enough to enhance our own. Like my grandfather's.

Morris Abrams Inc.

It was New Year's Eve, 1899, and the vibrant capital of New World commerce and culture was decked out and lit up. As his ship edged into New York's harbor, nine-year-old Russian immigrant Morris Abrams was spellbound by the spectacle before him. Fireworks blazed across the night sky. It felt like it was all for him. He instantly fell in love with America.

The flames of my grandfather's lifelong passion and optimistic embrace of this country were never doused by his hard childhood, two world wars, the Great Depression, and several recessions.

In his early teens, Morris went to work for a hardware merchant named Lemkin on Center Street in lower Manhattan. He learned the business and helped his boss build it. He married, and in the early 1920s he and his wife had three young kids and an apartment in Brooklyn. Lemkin was nearing retirement. When he refused Morris a partnership and prepared to pass the business on to his own son, Morris decided it was time to set up shop for himself.

Just down the street, at 196 Center, Morris Abrams Inc. was established in 1922 to supply hardware, machinery, and equipment to manufacturers, retailers, and contractors. All went well, and when my grandfather took in an equal partner named Morris Zippert in the late '20s, the young company was thriving. The two partners expanded and hired friends, family members, and others, and the company grew until the Depression hit. In late 1931, they were hanging on by a thread. They considered their options. With their heads still above water, they could close and walk away owing nothing. If they stayed open and things kept on as they were, they might be forced into bankruptcy, an uncomfortable option because they didn't want others to get hurt. It was a tough call. They decided to hold on for another few months, hoping that Roosevelt's inauguration would make a difference.

It did. As soon as the new president occupied the White House, old customers started returning. Lamp manufacturers, ventilation contractors, woodworking shops, tool and die makers—all became active, commerce began to hum, and doors began to open. The business grew slowly over the next decade. The U.S wartime effort in the 1940s gave it a big boost. By 1950, the payroll had swelled to 110 employees.

They moved to larger quarters at 90 Hudson Street, a solid, dignified seven-story building that occupied most of a city block and was capped by eighteen great arched windows on the top floor. The Morris Abrams Inc. clothbound catalog from 1955 is black with embossed red and white lettering. Included in its 569 pages are more than a hundred pages of woodworking and metalworking machinery: lathe grinders, milling machines, punch presses, power shears, benders, the full line of Delta woodworking tools, Crescent planers and band saws, Linley jig boring machines, etchers and engravers, blast furnaces and bench furnaces, and welders of all kinds. Each tool is presented with high-quality photographs supported by carefully worded text and detailed pricing. Then comes every kind of electric and pneumatic portable tool, followed by cutting tools, abrasives, taps, dies, blades, bits, heads, and reamers. One hundred and ninety-one different varieties of hand files, scrapers, rasps, and rifflers are offered in different sizes, at prices ranging up from $3.00 a dozen. There are sixty-five pages of micrometers, calipers, gauges, dial indicators, and other precision measuring tools, mostly made by Starrett Company and Brown and Sharpe, two venerable New England manufacturers that date back to the nineteenth century and still exist today, their product lines virtually unchanged. There's no sign of Makita, Hitachi, or Bosch—everything in this catalog is made in the U.S.A. Every imaginable hand tool is here—hammers, saws, clamps, pliers, shears, snips, cutters, screwdrivers, and planes. The catalog ends with a hardware section that has toggle bolts, cotter pins, woodruff keys, washers, screws, nuts and bolts, and a miscellaneous equipment section that has grease, graphite, and gear lubricant mixed with spring winders, belt lacers, and spray painters. From the everyday to the arcane, it's all there.

On page 195, one of the many toolboxes shown is the Union Super-Steel Cantilever tool chest, listed at $5.60. This is the one I'd see beside my grandfather's desk when my father dropped me off at the store on visits from our home in San Francisco. I'd sit in the office while we talked for a while. Once my grandfather had been brought up to date, he would hand me the brand-new toolbox and tell me, "Go fill it up, have fun, and behave yourself."

As I wandered through the maze of storerooms that felt like caves filled with treasure, the workers helped me find the stuff I wanted—

braces and bits, block planes, those wonderful ratcheting Yankee screw-drivers, and boxes of fasteners that fit in the neatly divided toolbox trays. Each object on the shelves was wrapped in oilcloth and individually boxed, and when I was through the tool chest was like a bag of presents. I hauled it to the airport when we left, put it on the DC-7 for the long flight home, and had lost pretty much everything in it by the time we returned two years later for the next round of this ritual.

Morris Zippert's son, Ira, had come into the business after he returned from the war. But none of my grandfather's three children was inter-ested. My aunt became a biologist, my uncle an actor, and my father, the youngest, a physician. No hardware merchants in this bunch. They took to heart their father's message about the value of education, and their motivation carried them to their different fields of interest. How did my

The 1955 catalog of Morris Abrams Inc. (Photo by Randi Baird.)

grandfather feel about having nobody to carry on the business that bore his name?

During the 1950s the company spawned many others. Senior sales-people left and took lucrative accounts with them. One took the little International Business Machine account and formed Bruckner Supply, which, after some years as a general industrial supplier, began to concentrate on supplying the computer industry. Bruckner became successful and was recently sold to a larger company for $125 million. For Morris Abrams Inc., however, the 1960s brought declining fortunes. The loss of good people, and the business that left with them, took its toll. The company moved to Long Island, closing operations on Hudson Street. They began to specialize, supplying school woodworking and metalworking shops. In the '60s they sold the company to Power Pipe and Supply and it continued to spiral downhill. My grandfather kept coming in for years, but there was little for him to do except collect overdue balances. The company was sold several more times, and the final owner let Ira Zippert go, sold the Morris Abrams name, and went bankrupt. The name still exists, like a shadow, and is sometimes used in bidding for New York City municipal contracts.

My memories of the business will always be as I saw it on my child-hood visits, during its heyday on Hudson Street. Built on solid foundations of honesty, respect, and loyalty, it seemed to be, at once, stable and magnificent. Did it break my grandfather's heart to see the decay of the institution he built? Nobody seems to know for sure. He died just before his one hundredth birthday (when she heard about his death, my wife said, "That can't be—he's too old to die") without conveying those feelings. He never liked to dwell on the past, but I imagine he must have been hurt by the progression of events. The business could have continued to thrive and endure. But just as Lemkin's legacy plans did not include my grandfather, my grandfather and his partner did not make sufficient provision for the people who helped to build their business. Their employees left, each time taking a piece of the business with them and contributing to the disassembly of the enterprise.

The Company as Community

I have come, through my own experiences in life and business, to think that small companies like my grandfather's do not have to die this kind of death. Is such dismantling the inevitable consequence of what economist Joseph Schumpeter termed the "creative destruction"[4] of free-market forces? Or is it a symptom of one of contemporary America's legacies, the frontier mentality (when we are finished with something, we throw it away, we move on, we make a fresh start)? I realize now that the demise of my grandfather's business is, for me at least, heartbreaking.

When businesses like my grandfather's are cast aside or wither and fade away, we testify neither to inevitable machinations of the market-place nor to some grand impulse of the pioneering spirit, but only to our own lack of common purpose. The loss is more than economic. Social capital that no catalog or chart of accounts can itemize is squandered. Thousands of businesses dying this kind of death in thousands of communities may be as damaging to the social fabric as the loss of family farms or the breakdown of marriage.

I recognize that not every company can be, or should be, viewed as the starting point for the establishment of a permanent institution. What I question is the lack of value placed on maintenance of those business communities we create and long-term commitment to the communities within which those companies operate. I question the unfettered allegiance to an economy whose rewards have become skewed toward the distant and the global at the expense of the local, and whose system of incentives encourages wage servitude and environmental recklessness.

I come to these conclusions thirty years into my business career, as the cofounder of a small company that now has $6 million in sales, sixteen owners among its thirty employees, and a dedication to the kind of enduring connection to community and place that we hope will survive us and, even, our children. We are still very much a work in progress, but we have set out—briskly, although not unerringly—along a path to a more democratic, more responsible, more permanent kind of company.

The Cornerstones

Our attempt to remake the way we do business pivots on eight essential cornerstones that emerged as I looked back over the arc of our collective past. I did not predetermine this number or decide to urge these ideas to be cornerstones; instead, I discovered them during the course of my endeavors, picking through them like a mason building a wall. The stone-mason sorts through a pile of material to find just the right stones for the base, the corners, the fillers, the stretchers that lock the wall together, and the capstones to finish it off. He discovers the wall as he builds it, as I found the cornerstones of our business. Each cornerstone revealed itself as I sorted through the pile. Once found, each could be placed, and together they became the foundation of a structure that is perhaps crude, but sturdy.

This book is an account of finding and setting these cornerstones. It's the story of the development of a particular design for business that is built on these underpinnings:

- cultivating workplace democracy
- challenging the gospel of growth
- balancing multiple bottom lines
- committing to the business of place
- celebrating the spirit of craft
- advancing "people conservation"
- practicing community entrepreneurism
- thinking like cathedral builders

These principles and practices are embedded in our company. The values I sensed in the way my grandfather did business—honesty, integrity, loy-alty, and optimism—are at the core of this company, too, but they are lodged in a different context.

Cultivating Workplace Democracy
In 1987 we restructured South Mountain from a sole proprietorship to an employee-owned cooperative corporation. It was an important moment in the history of the company. Ownership has become available to all

employees, enabling people to own and guide their workplace. The responsibility, the power, and the profits all belong to the group of owners. In her book *The Divine Right of Capital*, Marjorie Kelly, publisher of *Business Ethics*, says:

> Thomas Paine's vision was of "every man a proprietor." It's a worthy ideal, to own one's place of work. But in the corporate era, most citizens are necessarily employees, and always will be. We need a new economic vision for a new era: not every man a proprietor, but every employee an owner.[5]

Shared ownership and control is our method at South Mountain. Every employee an owner is our intention. More than half of our thirty employees are full owners; each time another comes in, and each time a new management invention encourages more voices to be heard, we move steadily toward the goals of democracy, fairness, and transparency.

Challenging the Gospel of Growth
A cherished business doctrine is that growth must be a primary business purpose: "grow or perish" is a mostly unquestioned truth. At South Mountain we favor certain kinds of growth, but not expansion for its own sake, which author Edward Abbey described as "the ideology of the cancer cell."[6] We embrace growth to achieve specific goals, but always with consideration of the consequences: it may disrupt and endanger treasured qualities. We look for ways to develop and thrive without enlarging, thereby holding to limited growth. When we grow, it is by intention rather than in response to demand. We think about "enough" rather than "more"—enough profits to retain and share, enough compensation for all, enough health and well-being, enough time to give our work the attention it deserves, enough communication, enough to manage, enough headaches.

Balancing Multiple Bottom Lines
We assign priority to a collection of bottom lines while consigning the traditional single bottom line—profit—to its appropriate role as a vital tool that serves the others. We try to satisfy the bottom lines of being

glad to come to work, making a good living for our families, meeting the expectations of our clients and business associates, caring for the environment and one another, promoting fairness and health, and improving the world, one house at a time, one neighborhood at a time, one island at a time. We propose cooperation, rather than competition, as the avenue to business success. Metaphorically speaking, these are all bottom lines, even though they cannot necessarily be quantified or measured.

Committing to the Business of Place

We have a long-term investment in the small island community where we work. All our eggs are in this one geographical basket. People often ask why we would devote such effort to a place that is widely perceived to be just a playground for the wealthy. It surely is that, but it is many other things as well. With all its strengths and weaknesses, assets and problems—its status as a world-famous resort; the vast socioeconomic diversity and discrepancies in its population; the dwindling of its community of fishermen and farmers; the growth of a new, sometimes fragile community of islander and washashore craftspeople and artisans—this is the place that we know best, the place that must serve as a laboratory for our experiments with small business. It's a good place to model new solutions, because it's socially complex and it's still in fair condition. There's hope for it. We have committed to limiting our work to Martha's Vineyard, with the exception of educational work beyond our shores, and to doing everything that we can to make a difference in the quality of our community and the stability and balance of the local economy.

Celebrating the Spirit of Craft

In all that we do, craft is the essential unifying concept. Although the joy of my own work is in its variety—I meet, talk, plan, write, design, visit, advise, learn, manage, make coffee, and feed the scruffy stray cat that lives under our porch—the thing that inspires my work is the craftsmanship I see around me. Craftsmanship is, for many of our employees and employee-owners, the aspect of our work that evokes the most passion. There is a tremendous sense of accomplishment in making something beautiful, something that employs appropriate materials, tools, and techniques in good proportion and functions with grace and ease. The

1980: The Sagan house in construction. (Photo by author.)

Balinese have a saying: "We don't have any art. We do everything as well as we can." They know craft. This is the spirit we are pursuing.

Advancing People Conservation

The Vineyard has a serious affordable housing crisis. The island's captivating charm is culpable. It wasn't so long ago that young people could find a host of housing options, but now our housing resiliency is gone. Staggering increases in real estate prices and the high costs of island living have made it impossible for young and working people to afford homes here. The community is still intact, but it's endangered. People are bailing out. We're losing parts of the essential fiber of our community, and those who purchase the homes left behind do not fill the same important civic roles.

Peter Forbes of the Trust for Public Lands says, "The future of conservation and preservation will be determined not by how much land and how many buildings we set aside, but how many people who know and love the land we can help to stay there."[7]

The affordable housing story is about people conservation. It's a story about sustainability. People conservation is the essential complement to land conservation. Can our problem be fixed? We have decided, as a company, to invest heavily in the notion that it can.

Practicing Community Entrepreneurism

Making expensive homes (often second homes) in a resort community has many significant returns: freedom to explore craft, opportunities for the pursuit of quality, relationships with interesting people, and financial rewards. By itself, however, it does not directly serve a broad social purpose—beyond providing good jobs to those who do the work and good homes to a fortunate few—no matter how socially purposeful we are in the way we do it. We have attempted to address this reality by using the financial resources and the web of relationships that derive from the work to help solve community problems and to encourage a better future for the place where we live and work. We bring an entrepreneurial approach to these efforts, taking risks and learning from both our public failures and our small successes.

Thinking Like Cathedral Builders

Our view of time is squarely at odds with short-term business thinking. The work of South Mountain Company will not be finished in our lifetimes; it will continue for generations. British business philosopher Charles Handy gives perspective:

> Cathedrals inspire. It is not only their grandeur or splendor, but the thought that they often took more than fifty years to build. Those who designed them, those who first worked on them, knew for certain that they would never see them finished. They knew only that they were creating something glorious which would stand for centuries, long after their own names had been forgotten. They had their own dream of the sublime and of immortality. We may not need any more cathedrals but we do need cathedral thinkers, people who can think beyond their own lifetimes.[8]

We try to think about our work as the cathedral builders thought about theirs. We try to think for generations, as we try to build for generations. We recognize that any wealth that has accrued to us has been provided in significant measure from the capital of this community, and it is our obligation to ensure that our business becomes an enduring part of the fabric of this community. To make a durable, robust, and flexible business community that outlasts its original owners, we plan for succession, so that as we age we can gracefully depart and leave the company—vibrant, stable, learning—in the hands of others.

Each of the cornerstones will be discussed in detail in its own chapter, but the principles are interdependent and often intersect, and they are therefore woven into the fabric of the whole. I will tell how we came to our beliefs and practices, how we transitioned to employee ownership, how the tenets of our practice emerged, and how we came, with our share of fumbling and bumbling, but with growing confidence, to be who we are today. I will examine the significance and meaning of our journey and consider where these efforts might lead and what new endeavors they may support.

1981: The new office frame rises behind the flowers and the '53 Pontiac. (Photo by author.)

The Slightness of Our Knowledge

The process of making the book has turned out to be surprisingly like that of making a house. You design the book and then you build it. You have to conceptualize it (design it), write it (build it), edit (move the walls that don't work), and acknowledge (honor the many relationships that combined to make it happen). Then you have to let it go.

Partly this project is an outpouring of gratitude for the things that have gone our way, the people whose paths we've crossed, the pure dumb luck that has allowed us to practice the craft we love, and the splendid group of colleagues and fellow owners who have been company on this journey, which has been, for me, more fun than I could possibly have imagined. And partly the purpose is the hope that a few of our chance discoveries will be helpful to others.

The question at hand is this: Can small business, supported by strong underlying principles, help make better lives and better communities? To go farther, and perhaps too far, can business conducted this way help us be kinder to ourselves and to one another, to the planet, and especially to our children? Is it a stretch to say that the more fully we are fulfilled in our work, the more fully we can love both our children and our community? And that the more fulfilled we are, the more we can help build a future that's sane and just? If I overreach, it is only my enthusiasm for the possibility that is at fault.

Here, then, is one small business on one small island. Its lessons emerge from its story, but it is only one of many stories of small-business experimentation that are unfolding today in the wealthiest nation the world has ever known. The simultaneous development of countless such stories, rich with intersecting themes—like so many shingles on a steeply pitched roof that provide shelter only because they are woven together—is what gives added meaning to any one account. Recognizing this, I offer our story so that it may take its place among the others, and perhaps, combined, they might together alter, in some small way, the chemistry of our culture.

My good friend Lee Halprin says about writing:

> All truth knows only a little of what is. And all writing—all true writing, even—names only a little of what's true, even. Now, some

writing somehow knows that, and somehow shows that. But certainly a lot of writing does not show that it knows the *slightness of its knowledge*. And lots may not even *know* the slightness of its knowledge. I think it is good to consider—quite independently of any particular writing task—how one feels about this relation between what one can say and what one knows and between what one knows and what one doesn't, and between what is known and what isn't, and between what's knowable and what isn't. I think it's good to think hard enough about this for the thought to somehow color one's writing, somehow to seep into it.[9]

That feels particularly right to me, but hard, too, to uphold. I'll be trying to remember the slightness of my own knowledge as I proceed, and trying to separate what I think I know from what I'm certain I don't. What I think I know is that after thirty years we are only at the beginning of this story. I cannot know where it will ultimately lead. I hope the process of telling this story will help me know more about where we have been and where we are headed, in the same way that my colleagues at South Mountain Company continue to help me know more, every day, about the community we build and the company we keep.

South Mountain Company 2004. (Photo by Bob Gothard.)

CULTIVATING WORKPLACE DEMOCRACY

$\lceil 2 \rceil$

Through the 1970s South Mountain was truly an extended family business. My wife Chris and Mitchell's wife-to-be Clarissa were integral parts; they plastered, painted, set tile, and stacked lumber. Chris prepared the bills and trucked materials from off-island, and we all lived together at the run-down farm Clarissa had recently inherited. Our few employees became close friends and hung around the farm as well. As we hired more employees and Chris and Clarissa began to work "in town" as teachers, their role in the company diminished. The business expanded, and we built a new office next to the old shop.

By the early 1980s Mitchell had developed a strong appetite for farming; he gradually distanced himself from the business and we formally disbanded our partnership. Chris and I built a house on an adjacent piece of land that Clarissa carved off for us, but the business remained in its old quarters at the farm.

One night in 1984, as we slept in our new house up the hill, Chris and I were awakened by a friend.

"John, wake up! Wake up! The shop's on fire!"

I jumped out of bed and into my clothes, bolted out the door, and ran down the hill. The sky was bright with flames. The shop was engulfed in fire. Firefighters had their hoses aimed at our precious little office, next to the shop, which was smoking but not yet burning. They managed to save it—what a gift—but the shop burned to the ground. For the next few days we cleaned up the wreckage and hauled truckload after truckload of twisted metal to the dump. It was a gloomy scene; all our painstakingly collected and restored old hand tools and cast-iron machines, and that hallowed barn, were gone.

1983: End of the workday with "midlife painters" in the foreground.

We regrouped soon after. Chris and I mortgaged our house, and with that money the people in the company began building a new shop on our property. Some months later South Mountain moved into a well-equipped shop with spacious offices above. The newness was strange, but the space felt good. We were back on track. The dissolution of the partnership with Mitchell was complete. I was left as the sole proprietor of our common creation, which was now ten years old.

I felt uncertain in this new place, but curious and engaged.

The Apples in a Seed

We had become committed to the idea of doing both the designing and the building. We took our cues from accounts of the old master builders of the Middle Ages, the pioneers of early America, the Arts and Crafts movement, and the Shakers. Our abilities were rudimentary and our aspirations high. We were devoted to fine woodwork and to alternatives to conventional construction practices. We combined timber framing, passive solar, and an eclectic, unschooled design sense to make learn-by-doing buildings. We had mixed success. With no formal training and little experience, we were unconstrained by knowing what couldn't be done (and conversely unaware of much that could) and equally unencumbered by skill. We unnecessarily reinvented the wheel regularly. Historian Daniel Boorstin, in his book *The Americans: The National Experience*, speaks about the development of the American factory system, which has parallels to our own experience:

> The system, which later was to have the look of grand invention and bold discovery, began in the casual experiments of men encumbered by century-accumulated skills and intricate social regulations. If [it] was a triumph of organization and of cooperation, it was also a triumph of naiveté, for its essence was a loosening of habits and ways of thinking. Ignorance and "backwardness" had kept Americans out of the old grooves. Important innovations were made simply because Americans did not know any better.[1]

1985: Following the fire we tackled the new shop and office and moved in soon afterward. (Photo by author.)

Our minor successes came in similar ways. We didn't know any better.

I had a vague sense that we were developing something of value, but I couldn't yet articulate what our successes might suggest. We were learning at breakneck speed and that was enough. At the tail end of the 1970s a series of important projects began to shape our future. I stopped doing carpentry and concentrated on design and project management. We built two projects that we didn't design, which confirmed our devotion to the design/build integration (since then we have never veered from that course). Finally, in association with the nonprofit Energy Resource Group (which we, with others, had helped to form in order to promote renewable energy), we orchestrated the research, design, and construction of a solar greenhouse attached to the Edgartown School. Teachers, students, and members of the public participated in the barn-raising-style event. It was our first significant venture into community demonstration work—a harbinger of things to come.

At the time there were ten employees, including two who had been with us nearly from the beginning. I loved the work and I loved the people. We began to think about the company as an entity and to conceive of our role in the community. We learned land-use and building techniques that embraced environmental concerns. Our affordable housing projects balanced our high-end work.

These early efforts were beginning to lead somewhere, to germinate. Author Ken Kesey once said, "You can count the seeds in an apple, but you can't count the apples in a seed." That's how it felt to us.

But while the philosophical underpinnings were evolving, we were experiencing unexpected growth and loss of intimacy within the company. The business seemed fragmented and adrift. We were flat-out busy—a perpetual motion machine—and our core became uncertain. Our identity as a family business working out of Allen Farm was gone. So who were we now? I had no idea.

A Democratic Workplace

Our growth, although hardly explosive, had nevertheless brought us to an unsettling perch, as if we were leaning against a wobbly railing on a second-floor balcony. No longer could we run solely on intuition and gut; the business had become too complex. We needed a system that would allow the familial qualities we cherished to be maintained in a larger context. Issues that had not been evident became visible and urgent. The two employees who had been with us almost from the beginning, Steve Sinnett and Pete Ives, came to me and said they wanted to stay with South Mountain, that they wanted to make their careers here, but they needed a greater stake and more than an hourly wage. It was time to do more than reinforce the railing.

Steve had been our first employee and had become a close friend. He had migrated to the Vineyard after college, attracted by the waterfront scene, and had worked as a crew member on the schooner *Shenandoah*. He soon became a fixture at the farm and in the company. His personal qualities—his restless desire for all things to be better for all people, his intense loyalty, his unflagging team spirit, and his pitch-in-when-the-going-gets-rough approach—combined to make him indispensable.

In 1978 Pete Ives had come to work. He was an accomplished mason, painter, drywaller, floor sander, tile-setter, and surfer, but he had never done a lick of carpentry. He was hungry to learn and eager to pitch in. Once he said to me, "Just tell me what to do. I'll do anything you ask, as long as I don't have to tell anyone else what to do." He was dedicated, versatile, and talented and thought that he had no leadership qualities. He became a superb carpenter in a very short time. He began to find confidence in his work. He learned to be a foreman, first reluctantly, then with pride.

The three of us put our heads together and decided that the situation with the two of them was not likely to be unique; it would come up again and again if the company continued to succeed. How could we remedy the current circumstance but also welcome others, in the future, to a new status that offered more participation in decision making, greater responsibility, and opportunities to share profits? The journey of inquiry and experimentation that would follow led to our discovery of the first cornerstone of our business: employee ownership and control to create a democratic workplace.

Our first company Christmas card photo in 1986—it's been an institution ever since.

Adjusting the Model

I said I wouldn't concentrate on the troubles of our nation's economic system, and I won't, but I need to say a little in this regard. I agree with those who believe that too many people in today's economy lack a sufficient voice or stake in decisions that affect their lives. The consequences are serious. Author William Greider is convinced that most Americans think something is wrong "in the contours of their supposed prosperity." In *The Soul of Capitalism*, he writes:

> I do not find these complaints restricted to the poor or struggling working-class, though their struggles are obviously more stark and often desperate. . . . I have heard people from nearly every income level express an oddly similar sense of confinement, as if their lives were trapped by the "good times" rather than liberated. . . . Think of the paradox as enormous and without precedent in history: a fabulously wealthy nation in which plentiful abundance may also impoverish our lives.[2]

Our wealth has come at considerable social and environmental costs. Unless we provide a greater stake in economic decision making for more people, those costs are likely to continue to increase.

In *The Divine Right of Capital*, Marjorie Kelly expresses the view that there are significant opportunities to make our economic system more democratic. She says this may seem daunting when we consider the power of the financial elite, but:

> . . . we should remember that the power of kings was once as great. The very idea of monarchy once seemed eternal and divine, until a tiny band of revolutionaries in America dared to stand up and speak of equality. They created an unlikely and visionary new form of government, which today has spread around the world. And the power of kings can now be measured in a thimble.[3]

She makes the point that democracy has been an unstoppable historical force, and that if it "hasn't stopped at the doors of kings, it is not likely to stop at the door of financial aristocracy."[4]

I was thinking along these lines—although without the rich historical sense of Greider or Kelly—as I considered my business options in 1986. I wanted the people in our company to feel prosperous and fulfilled in their work. I decided, with Steve and Pete, to investigate structures that would distribute both ownership and control. Greider compares employee ownership to homesteading. Instead of the government making land available, as in the past, he advocates low-cost capital that workers could borrow from the government to purchase equity in their companies. They would repay their debt from their share of the profits, like "homesteaders 'earned' their property by farming and ranching on the land."[5] For us there was no need to wait for government.

This was surely a hinge point for South Mountain. For a while I was *unhinged*—alternately frightened and excited by our deliberations. I had the power, and the greatest financial and emotional investment; therefore, I had the most to lose. Under my ownership the company had become a viable, profitable entity with a strong reputation and a backlog of work. Sometimes, during those sessions, it felt like control was slipping away, like I was tugging on the reins of a runaway horse. Then it occurred to me that perhaps I had the most to gain: aside from the lure of clearing this new path and seeing where it led, the possibility of shared responsibility and ownership promised new freedoms for me and new achievements for the company.

Our inquiries led us to the concept of a worker-owned cooperative corporation. It seemed radical but promising, especially if we could make the shift to employee ownership and control in a gradual, carefully measured way. Expressions from the participants—excerpted from the notes of those early meetings—evoke the tone of the discourse:

"Our structure should guarantee that anyone who makes a career here should be extended the privileges, responsibilities, rewards, and headaches of ownership."

"The underlying premise for any change we make must be mutual respect and trust. To lose what we've created in that regard would be tragic."

"We are a small, successful company with a strong reputation and track record but an informal, relaxed structure. Very little about our governance and performance systems is defined except by habit, experience, and our various quirky personalities."

"Pete and Steve have put a lot into [South Mountain Company]; the restructuring should reward them without taking from John, who has led us this far, and the new structure should gradually reward others who make similar commitments."[6]

With some trepidation, we hired Peter Pitegoff, an attorney at the Industrial Cooperatives Association, now known as the ICA Group, to advise us. I worried that hiring ICA would mean there could be no turning back. The safety and insular quality of sole proprietorship, which I had only recently earned, was about to be cast off. Engaging Peter was a semipublic announcement of intention. As a lifelong skier, I compare the feeling to summoning the nerve to drop into a steep couloir when you can't see below the crest and you know your skiing buddies are down below, waiting for you to come. With tips pointed down, I pushed off gingerly.

At a meeting in May of 1986 I expressed the view that Chris and I—who owned and lived on the land on which the South Mountain Company premises were located at that time—needed to retain control of the property. I was also concerned that, in the business, my customary freedom to act solo might become so constrained by shared ownership that I would no longer be comfortable there. It was the first ringing of the bell that all who make the shift from sole proprietors to employee-owners must hear. To say it another way, what if the thing we've built with painstaking care evolves into something we don't like? It's a serious risk.

Meanwhile, it was suggested at the same meeting that "at the beginning John could have veto power over new owners, jobs we take, hiring and firing, and wages." I read that now and chuckle to myself, "Hey, what else is there?" By that arrangement I would essentially have been keeping *most* of the control. The idea was to spread that control widely, so the voices of all the owners had meaning. You can't steal second without taking your foot off first, and we came to agree that the only protection needed was veto power over issues directly related to the property.

At Peter's suggestion we adopted a democratic ownership structure patterned after that of Mondragon, a remarkably successful network of worker-owned cooperatives in the Basque region of Spain. Mondragon has operated for nearly fifty years on the principles of employees as owners, labor controlling the enterprise and sharing the wealth, members participating in business management and decision making, a limited ratio between top and bottom pay, and education as the key to career development and progress. We made adjustments to this model to fit our own idiosyncratic needs as an organization converting to, rather than starting with, employee ownership. Particularly important was the institution of a lengthy five-year trial period before ownership for employees. This ensured a gradual transition, allowed time to measure commitment and suitability before people became owners, and provided room for training and building understanding before employees were thrust into policy decision-making. We established an ownership buy-in fee. We decided that this needed to be significant but affordable. If it was too steep it would discourage participation, so we set it at the price of a good used car, an expense everyone seems to be able to manage when necessary. The fee has increased slowly; at this point it's an uncommonly good investment for new owners and it's still equivalent to the price of a good used car.

Peter established a method for valuing and buying out my interest, drafted a set of bylaws, and developed a legal agenda for reorganization that laid out the process coherently.[7]

Restructuring South Mountain Company

On January 1, 1987, I transferred the ownership of South Mountain Company to a new worker-owned cooperative corporation. Steve, Pete, and I were the original three owners. Our jobs didn't change—I remained the general manager and Steve and Pete remained foremen—but in our new roles as employee-owners, our responsibilities did. My compensation for selling the company was in the form of preferred shares, which were converted to cash over a period of five years, and a full ownership share (a more detailed explanation of the mechanics of our structure can be found in appendix 1). The first meeting of the board of directors of the newly reorganized company convened on January 9, 1987. Attending were Steve, Pete, Peter Rodegast (soon to be the fourth owner), and myself. There were seven other employees at the time of the restructuring and they were all on a track toward ownership. This was a critical transformation in the life of the company, the setting of that first cornerstone of our developing business model. The full implications of what we were doing were not yet clear to us.

In 1989 Peter Rodegast became our fourth owner. Peter had been hired in 1983. He had studied architecture and had experience in both building and design. Over the years his presence in the design studio had become indispensable and he no longer was able to fit in much carpentry, but to this day, when asked at annual evaluations if he'd like anything about his job to change, one of his refrains is "Well, I wouldn't mind pounding some nails at the job site." Mike Drezner, a former teacher turned carpenter who was a bit older than the rest of us, was next in line, becoming an employee-owner in 1990. But all was not peaches and cream.

Within two years of the restructuring Steve—one of the original inspirations for it—left the company. He was an effective owner, but he was involved in many causes. Each was more worthy than the next, but at the time he wasn't able to balance them (and new parenthood for good measure) with his responsibilities within the company. The distractions caused a crisis on the job. Due to his longevity and take-charge approach, he was one of our foremen, but his projects were suffering and the people on his crew began to complain. We anguished together and decided, with his concurrence, that he'd be better off trying something else. His departure was disheartening and difficult, but amicable and successful. We owe Steve tremendous gratitude for his important contributions to the early days of South Mountain. After he left, he and a partner founded a landscaping company, a design/build company much like our own. That company, Indigo Farm, has become a kind of sister company, doing most of our site planning, landscape design, and landscape construction for the past several decades. Eventually they, too, converted to cooperative ownership. They continue to thrive, and we continue to collaborate, although Steve is now engaged in other work (most recently helping to rebuild a temple in Mongolia!).

Our sixth owner was Vicki Romanauskas, our office manager, who signed on in 1991. In 1995 she became the second owner to leave. She left under very different circumstances; her departure was simply a time to celebrate her decade of employment, the contributions she had made, and her upcoming marriage, which was pulling her away from the Vineyard. During her four and a half years of ownership she had accumulated sufficient equity that she would depart with a significant nest egg.

Steve and I in the '80s, fastening a guy wire for a wind turbine tower.

In the decade since, no owner has left. In that time we have gained twelve new owners for a current total of sixteen. Several more will soon qualify. Incorporating new members has worked well; all take the responsibility seriously and quickly become contributors. When we hire someone new, we assume that they will become an owner in five years. This makes us think differently about who we are hiring and why. The selection of new employees is as important as the structure. The five years is, in effect, a trial period. During that time our personnel committee clarifies, before each individual reaches eligibility, whether the employee wishes to accept the responsibility and whether the current owners wish to accept this person as a partner. An employee who makes it through five years without extenuating circumstances is likely to become an owner.

But ownership is not a requirement. Neither is it a right. It is a privilege to be enjoyed by those for whom it is appropriate and desirable. Some have remained here, working full time, as long as ten years without becoming owners, but this is rare. Ownership just isn't for them, or so they think; some of us argue differently and encourage them to join. We have noticed that since employee ownership is an integral part of the company culture, those who remain nonowners for a long time are, in subtle ways, isolated from important internal dynamics of the company, along with missing out on the opportunity to build valuable equity.

Prospective owners are expected to have three principal criteria: an understanding and intent that employment at South Mountain will be their primary work for the foreseeable future; a demonstrated ability, from the evaluation process, to work effectively and cooperatively; and a commitment to understanding and honoring the company's core values of quality work, ethical business conduct, environmental responsibility, and concern for others—in short, we expect that a new owner will be a good representative of the company.

The employee-owners are the board of directors of the company and they make all policy decisions. *Only* employees may serve on the board. Ownership is inextricably tied to employment; upon termination of employment or retirement, an owner's share must be sold back to the corporation. Although we get plenty of advice from accountants, attorneys, bankers, and consultants—the best ones we can find—and we have the utmost respect for them and listen carefully to their counsel, they do not make decisions about the company; only employee-owners do.

The key distinction that requires constant examination is the difference between ownership and management. The board sets policy. Management carries it out. The board has responsibility for issues that affect the future of the company, such as accepting new owners; significant personnel issues; compensation and benefits policies; profit sharing; general direction in terms of future projects and work; major purchases, investments, and expansions; company growth; new ventures; involvement in community projects; and major donations of time and money. It is not always clear whether a decision should be made by management or brought to the board. We are continually testing these boundaries and refining our process. Informal adjustments and new understandings are crafted over time.

All decisions of the board of directors are made by consensus, but there is a backup voting mechanism for those rare instances when we are unable to reach consensus (see appendix 2, "Meeting Facilitation and Consensus Decision Making"). Each owner has one vote. In eighteen years we have only had to vote three times. The first time we voted, many years ago, was over a trivial matter that provoked strong disagreement. We couldn't reach consensus, and finally someone said, "We're spending far more time on this than it's worth—let's take a vote." Without resolving our disagreement we voted and moved on. Easy enough. The second and third votes were about substantive questions; the dialogues were rife with conflicting views and debate. On each occasion we had several long meetings before determining that we could not reach consensus. The ensuing votes were not close, however, and there was no discernible rancor in the aftermath because the airing had been so complete. We have found the backup voting provision to be essential to the effective use of consensus decision making.

Everyone accepts the importance of balancing participation and efficiency. We know that since there is no map to guide us, we need to be comfortable with trial and error and nimble enough to alter the process as needed. Roy Morrison's title for his book about the Mondragon cooperatives in Spain, *We Build the Road as We Travel*, is an apt description of our ownership and management systems. We, too, are building the road as we travel.

What Ownership Means

We have been blessed with a congenial, respectful, thoughtful group of owners; there are few difficult struggles. As the ownership pool expands, there is more diversity, which leads to more disagreements but richer discussions and more thorough investigations. Monthly board meetings are informal, intimate, and humorous, but they are carefully facilitated; all have a voice, but the discussions don't wander or stray too far off topic. The average length of the board meeting is two hours. As a friend of mine says, "Anything longer than two hours is a workshop, not a meeting." Understanding that facilitation skill is essential to good meetings, even of small committees, we provide internal training and encourage the development of artful facilitation practice. We have found that knowledge of facilitation technique produces better meeting *participants* as well as better meeting *leaders*. When people learn what it takes to run a good meeting, they gain greater understanding of how to successfully collaborate.

As the company has grown, we have developed more collaborative management processes to go along with our shared policymaking. Much of our management work is done by committees. An executive committee, a production committee, a design committee, and a personnel committee meet and act on a regular basis. A charitable contributions committee, an education committee, and a facilities committee meet as necessary. The group of owners has the ultimate authority, but it delegates much of the trust and authority to management.

Ownership has different meaning to each of us. Here's a sampling of the opinions of South Mountain partners about the significance and spirit of employee ownership:

Peter Rodegast: "It's interesting to see how coworkers of all sorts evolve into business partners; it's different from just working with a few close friends. You end up with a diverse board and a variety of opinions."

Mike Drezner: "My notions of what ownership means have slowly changed as time has passed. Initially, I thought of it as analogous to [being] a shareholder: to have a voice in policy and business decisions but distance from personal matters within the company. I have come to learn that you do not decide the extent or nature of your involvement in company concerns. In essence, you become a parent to all South Mountain issues and your responsibility for the well-being of the company demands commitment. That understanding dictates a subtle change of perspective. You need to take what you know as an employee and use it only as a resource. The trick is to respond to South Mountain issues with honesty and objectivity and not to allow responses to be influenced by comradeship, personal financial gain, or the desire to avoid thorny interpersonal problems."

Peggy MacKenzie: "Since becoming an owner I have a deeper interest in just about everything that happens here. If I don't understand or agree with something I catch wind of, I ask questions so I get it, and I voice my opinion. Ownership is what each individual brings to it. I see it as a commitment as well as an opportunity. And frankly, it's a good investment, but somehow that's beside the point."

Derrill Bazzy: "Even before ownership I felt I was part of the decision-making process, because South Mountain has always been the kind of company where all employees have a voice. Everyone can choose to be involved to whatever degree they like. What's different about ownership is the financial aspect. Ownership offers me the opportunity to make a good investment, and to make it in a company I believe in, have a say in, and plan to remain a part of."

Kane Bennett: "What makes South Mountain special to me is the people who work here. When I was invited to become an owner I felt honored that they wanted me to join them and become a larger part of what they have worked so hard to build."

Phil Forest: "Sharing ownership of this company has given me a greater sense of community. . . . I feel empowered by the chance to help guide the company. I have increased my commitment, discovered more opportunities, assumed more responsibility and acquired more influence. . . . I belong here."

A feeling of belonging. Diverse views. Empowerment. More opportunity. Deeper connections. The system seems to encourage these. Our legal and financial covenants complete the framework within which the employee-owners share the wealth the company creates and control its destiny. This is not about a *sense* of ownership or a *sense* of control. Corey Rosen of the National Center for Employee Ownership once said that giving employees a "sense" of ownership is like giving them a "sense" of dinner.[8] This is the whole meal.

Sharing Profits

The financial opportunity is the appetizer, setting the table for the other courses. Personal equity is accumulated through profit sharing and is recorded in each owner's individual internal capital account. These are paper accounts (not cash accounts) backed up by the company's net worth. They begin with the membership fee that each owner pays and they grow after each profitable year. South Mountain has two forms of profit sharing: All employees—owners and nonowners alike—share roughly 35 percent of the profits each year as cash bonuses, based on hours worked during the calendar year (this partially mitigates the hierarchical wage scale; we have distributed roughly $750,000 this way over the past six years). What remains after these bonuses is net income. Of this, 50 percent is distributed into the owners' equity accounts. The other 50 percent remains in the business as retained earnings. The equity accounts measure the financial stake of each employee-owner.

Employee Ownership Questions

When we decided to restructure South Mountain in 1987, I mistakenly thought that what we were doing was more symbolic than substantive. It has been more meaningful and valuable than I could have imagined. It was an experiment that involved taking a major psychological leap as well as a legal one. I suspect that it has been as rewarding for others as it has been for me, and there's no question in my mind—although I can have no definitive proof—that it has been a critical factor in the long-term success of the company. But a few questions bear asking.

Why did our investigation lead us to a worker-owned cooperative instead of, for example, creating partnership positions for Steve and Pete?
A conventional partnership approach might have done the trick. It would

have served Steve and Pete's personal needs and been easier to create, but we wanted a long-term solution. We also were intrigued by the challenge of finding out to what degree sharing ownership and control could democratize and humanize the workplace. If we were going to go through all this trouble, we thought, it might as well be for a greater purpose. Our backgrounds and personal histories were also significant factors. Our countercultural bent, communal spirit, and political views caused us to imagine cooperative ownership to be an appropriate response to the issues we had identified. Naïveté probably played a part, too.

Is there a future for employee ownership in the mainstream of American business? There is evidence to suggest there may be. The most common form of employee ownership is the Employee Stock Ownership Plan (ESOP). An ESOP is, in fact, a pension plan, but it differs from conventional pension plans in several ways that make it a useful device with which employees can purchase all or part of the company that employs them. ESOPs invest in the company itself, rather than in outside companies, and ESOPs have the ability to borrow. John Logue and Jacquelyn Yates of the Ohio Employee Ownership Center recently conducted an extensive study of 165 ESOP companies in Ohio. In the book in which they analyze their findings, *The Real World of Employee Ownership*, they report that in 1998 there were roughly 11,400 ESOP firms nationally, with 8.5 million employees (about 8 percent of private sector employment) and $400 billion in assets held by the employees. Some of these, like United Airlines, Andersen Windows, Procter and Gamble, and Publix Supermarkets, are familiar names. Most are smaller and less visible.[9]

Unlike cooperative corporations, ESOPs tend to distribute only a portion of the ownership and do not necessarily extend employee participation and control in decision making. The Ohio study found that "almost three quarters of ESOPs actually do take some steps to broaden participation in firm management. The changes made in most of these firms are quite modest, however, and a quarter of firms make no changes at all."[10]

Peter Pitegoff, in a recent article titled "Worker Ownership in Enron's Wake—Revisiting a Community Development Tactic,"[11] suggests that this type of employee ownership was tarnished when Enron failed and thousands of employees lost the $1.3 billion that they had, together, invested in company stock. Worker ownership at Enron, of course, gave few employees a meaningful say in corporate affairs, but the fact that the company overstated earnings and top executives sold their stock while others couldn't contributed to public perception about abuse of employee stock ownership. In that case and many others, the relationship between managers and so-called employee-owners is more like that supposedly expressed by Louis XIV's tax advisor: "The idea is to pluck the goose in such a way that you get the most feathers with the least amount of hissing."

But ESOPs are a start, and Logue and Yates support their potential. They discuss incentives that might create more democratic enterprises and wider employee ownership. They point to the proposals of Representative Dana Rohrbacher (R-Calif.), a conservative Republican and former speechwriter for Ronald Reagan, who thinks federal tax incentives can be reshaped so that 30 percent of all U.S. corporations can be owned and *controlled* by their employees within a decade. Say Logue and Yates:

> Rohrbacher's employee ownership bill[,] first introduced in 1999, called for creating a new federal corporation type: Employee Owned and Controlled Corporation (EOCC). To qualify under the proposed EOCC rules, employees would own a controlling interest in the corporation, at least 90 percent of employees would be participants, and employees would have a right to vote their shares on a one-person, one-vote basis. . . .[12]

Is the concept of employee ownership growing and finding new advocates?
Logue and Yates found that in those companies in which greater employee participation is encouraged, there tends to be a corresponding jump in productivity and profits and a reduction in turnover. As this phenomenon becomes more widely understood, interest is increasing.

Cooperative ownership, which embodies more complete sharing of ownership and control, is growing only slowly. But it's a system that has been tried and tested over decades, even centuries. Pitegoff states:

Documentation of worker cooperatives in the United States dates back to the 1790s and continues through the Civil War era primarily in the context of organized labor. Striking carpenters in Philadelphia[,] for instance, formed a cooperative enterprise in 1791 to compete with local employers; and cabinet-makers, tailors, hatters, and saddlers there operated cooperatives in the 1830s.[13]

Pitegoff has noticed increased activity in worker cooperative development since the 1970s and notes that while they have never been any significant portion of the American economy, cooperatives have had a modest impact in certain industry sectors and regions and in certain historical periods. Industrial cooperatives were widespread in the late 1800s. More recently, according to John Curl, author of *The History of Work Cooperation in America*, "Cooperatives did about a third of the total farm production and marketing in the [United States] in 1980, with 7500 farmer co-ops and almost six million members."[14]

Traditionally, American unions have been suspicious of making workers into owners, with good reason. Giving workers stock is often part of a deliberate strategy to undercut unions by blurring the line between capital and labor. Said one union official, "When labor gets in bed with management, there will be two people screwing the worker, not one."[15] In a true cooperatively owned business like South Mountain, the employee-owners share both profits *and* control; there's no separation. The group is vested with both the benefits and the burdens of distributed power.

Why hasn't the socially responsible business community embraced shared ownership more fully?
The corporate social responsibility movement, which was exemplified in the 1990s by entrepreunerial companies like Ben and Jerry's, Patagonia, the Body Shop, and Stonyfield Farm, has begun to popularize ideas about commerce as a means of implementing people-centered human resources programs, promoting environmental sustainability, encouraging participation in local community life, and creating new forms of collaboration. But there's hardly a murmur, as far as I can tell, about distribution of ownership. Entrepreneurs are risk takers, but perhaps giving

up control seems like too great a risk to these pioneers who have already risked so much to build businesses that embody their personal values.

I've come to believe that giving up control is the business risk that has *the greatest potential to cause the greatest returns*. It's not unlike choosing to have a baby. There can't be anything we do in life more risky than having a baby, but for most people the perils are apparently outweighed by the potential pleasures and fulfillments. Not only did the loss of control seem to me a worthy gamble, but it felt like offering ownership *without* control would be like giving someone a car without turning over the keys.

Do employee-owned companies have difficulty raising capital?
Yes, often they do. The Mondragon model solved this problem by requiring 50 percent of retained earnings to be reinvested in the company. In this way, Mondragon avoids having to go to capital markets. The workers commit to reinvestment in the company, and capital remains tied to the community. Our system works the same way. Equity is assigned to each owner and we gradually build a reserve fund to buy out owners' equity shares when they retire, thus keeping the shares within the company.

Is it possible for a group of people, especially one that is forever expanding, to assume the full mantle of ownership? Can everyone truly engage? Can people with different occupations, orientations, and backgrounds have similar relations and depth of commitment to the thing they have agreed to share custody of?
In our case it hasn't happened quickly or without difficulty, but it has happened steadily. Degrees of involvement vary and the dynamics are constantly changing and evolving. It's no easy undertaking, and there are no guarantees of success. When success comes, it isn't likely to manifest in the ways we imagine. We are making fundamental changes in a complex human system.

Such a shift requires that we all learn new relationships, while keeping the company afloat. As we bring more people to the table, we create new potency and capacity that did not previously exist. Gradually participants come to understand that the only power is the power of conviction and expertise: to create change we have only to convince each other. It doesn't matter what you're trying to do—if you want to replace the flatbed truck with a pair of oxen, all you have to do is to convince the others that it makes sense. If we continue to use the truck, it's because we agree to, not because it has been predetermined or decided by one person that it's better than a team of oxen.

Tough Road

I have watched a number of other small companies in my field as they have made the change to employee ownership and control. The transitions are not always easy.

My good friend Merle Adams founded Big Timberworks, a wonderful company based near Bozeman, Montana, that designs and builds some of the most beautiful and inventive homes I've ever seen. A few years ago they adopted an ownership system not dissimilar to ours. Recently I asked him, "Looking back, Merle, four years later, what do you have to say?"

He replied, "It's been a rocky road, full of potholes. Wild animals have leapt in front of us, our tires have blown out, and we've slid into more than one ditch. Mostly there have been six guys in a car that'll hold only four, and some have stayed in and some have jumped out."

I was a bit taken aback by his negative assessment. "Whoa there, Merle, that doesn't sound so good. Say more. Knowing what you know now, maybe you'd go a different route if you had it to do over again, huh?"

He became more serious. "No. My life is better. The company is better. I'd do it differently. It all happened too fast. I could have extended the opportunity to just a few rather than many. And the shakeout that's happened is full of lessons. The 20 percent who weren't suited to ownership responsibilities cause disproportionate disruption. So we understand now that we need to hire and offer ownership opportunities more selectively."

It's important to consider the rate of change carefully. But when I spoke to Merle recently, he said that the turmoil had led to new stability, and that the company was in better shape than ever.

Last year I met with five young craftsmen who worked together in a small building company, Walnut Street Builders, in Burlington, Vermont. One of them, Dunbar, owned and ran the company, but they were all good friends and had worked together for years. They were interested in

Interior shot of one of the remarkable Big Timberworks houses in Montana. (Photo by James Salomon.)

restructuring as an employee-owned cooperative. Several forces were at work. Dunbar was tired of being the sole responsible person in a company of peers. Two of them were working as subcontractors because they were not interested in being employees, but they were valuable members of the team. Without a change, they were likely to go out on their own. The other two were not ready to go out on their own, but they were ready to get ready. None felt they would be satisfied long term to be just working for Dunbar. They all shared a desire to align their strong personal relationships with their work relationships.

In talking about the shift, Dunbar commented, "The people I'm interested in working with on a daily basis are my peers. The traditional model for a building contractor is not particularly suited to recognizing carpenters and contractors as peers. I had done my best to minimize the dichotomy within the model we had grown into, but I knew it needed a structural solution. As soon as we started down the seven-month road to reorganizing, there was a ground shift in attitude, effort, and vision. Clearly the new model was opening the way for the others to contribute on a whole new level. That's inspiring."

They restructured in August 2003 as Red House Building Inc. An interesting bit of evolution happened immediately after the restructuring. They were used to working on one or two jobs at a time, all five of them generally working together. As often happens in building, circumstance and schedule changes piled up several jobs at once. Then they were offered an exciting project that they did not have the capacity to take on but that they couldn't bear to turn down. So they geared up and hired six new employees. Suddenly the company had more than doubled and they had four projects going at once. Some among the original five (now owners) had to take on major new day-to-day duties immediately, in addition to their new ownership commitments.

I met several of them for lunch. I asked, "Given the stress of this fast growth and change, I assume you've been shoulder-to-the-wheel just getting through each week and you've hardly had a moment to register what's happened with the ownership, right?"

"Oh, no," Dunbar replied. "Without the new arrangement we never could have handled this. No way. I never would have taken it on alone. We're doing okay with it too. It's tough, but we're feeling our way and getting the job done."

I know what it's like when a small craft and service company suddenly doubles its workload—there are serious risks of quality, schedule, and financial failure. They say they were able to take this greater risk *because* of the new structure.

After twenty years in business, Graham Contracting, a successful design/build company specializing in high-quality residential and commercial renovations in the metropolitan Boston area, also recently restructured to employee ownership. They began with eighteen employee-owners, but soon after there was a shakeout period, like at Big Timberworks, and the number settled out at thirteen (there are seven additional not-yet-owner employees). Greg Graham and I spoke about the process. The departures after the restructuring had been not those of a bloc but, rather, random movings-on. He mused that the intense thinking about the future of the company stimulated equally intense personal introspection, and that some people had new inspirations about their work lives and, happily, pursued them. Others just didn't work well in the cooperative setting.

Soon after the ownership restructuring, Graham Contracting suffered a meaningful loss of revenue. The owners decided to forgo pay raises for the year. The thought had never occurred to Greg, who had considered annual raises to be a given. He was amazed at the reaction of his partners.

Greg had thought the hardest part was going to be letting go. But once it was done, he felt a sense of relief, and he finds that the troubles of letting go are being replaced by the excitement of seeing people evolve and realizing that his company will endure past his time. He feels that the shakeout filtered out the people who would have made it hardest for him to comfortably let go of the reins.

The Jury's Still Out

So far, employee ownership is working at South Mountain. If we consider the eighteen-year period between the restructuring and the writing of this book as an epoch in the company's history, the structure has withstood its first test of time. It has been easily adjusted as change has been needed.

One potential difficulty I see is a kind of cultural hardening of the arteries. There are occasions when I worry that we are becoming more concerned with security and less willing and able to take chances. Can we keep the conservatism that comes with age from overtaking our spirit of innovation? I don't know the answer, but I think that the goal is a healthy balance between stability and risk taking. The key to keeping vitality may be what business consultant Peter Barnes calls "intergenerational yielding." We, the owners and leaders of this company, must continue to welcome new leaders. We must look to younger people to do more than hold steady; the business must evolve if it is to continue to thrive. We're hard at work on this one, and we are beginning to have a broad spectrum of ages among our owners. Maintaining and expanding this diversity will become an essential task of the next decades, so that there will always be a fresh group to dare to innovate, and to carry on.

Another question: What happens as ownership continues to grow? Will we one day find ourselves with thirty or fifty people making policy decisions instead of sixteen? Will our current system work at larger scale? We cannot know.

I do not believe that restructuring to employee ownership will turn a business around. A healthy business and significant mutual trust are prerequisites. We must build businesses that are ready to take such a dramatic step. If we take a dysfunctional business and restructure, we can be sure we will have a dysfunctional worker-owned business when we are done.

Our governance system is a democracy with clear divisions of responsibilities and authorities. Much of the authority to act is delegated to management. This delegation comes easily, because this was the established mode of operation before the ownership was shared. The difference is that there is now a clear mechanism for discussion, debate, and change. The comfortable delegation of authority may be one of the advantages of a company *converting* to worker ownership and control rather than *starting* that way. The entrepreneurial leap of starting a new business has been achieved without constraints, and a viable company has been established. Restructuring becomes a part of the maturation process.

Even with multiple owners and voices we are rarely bogged down in process. Part of the reason for this is the acceptance of hierarchy based solely on expertise, not power. Most hierarchies serve two purposes: efficiency and maintenance of power. "Once the power aspect is gone," says Terry Mollner, founder of Trusteeship Institute, "people love hierarchy because of its efficiency, and they don't find it to be a barrier to healthy relationships with each other."[16] I think that's true in our case. Our decentralization efforts are not intended to reduce hierarchy; rather, they encourage more participation without decreased reliance on expertise.

We don't want the crew to sit around on the job and draw straws for who cases the windows and who builds the stairway. We want a capable supervisor who knows the people he or she is working with and assigns responsibilities appropriately. In the end, decisions must be made by those who have the expertise to make them. The decisions that get made on that basis reduce the number that must be made by consensus by the board of directors. We've been making decisions this way for the past eighteen years.

In thinking about the dynamics of employee ownership, I am reminded of the way the Roman army handled daily rations. Rations were in the form of large loaves of bread, each sufficient to feed two soldiers. This presented a problem, since when the soldiers had little to do, they tended to fight among themselves, particularly over who got the bigger half of the loaf. The Romans developed a nifty solution. They passed a regulation that one soldier had to divide the loaf and the other chose which half to take. Employee ownership is a similarly self-enforcing system. Each owner's actions on behalf of the others, and the company, are actions on his or her own behalf at the same time.

I hope I'm not overfreighting the ownership aspect. I understand that employee ownership is not the only way to encourage more responsible and more democratic business practices. But it does create fundamental differences when both control and profits accrue to the people who do the work. The apportionment of money in our current economic system is particularly vexing. How did it come to be that shareholders are the ones who accumulate most of the wealth? Charles Handy, in *The Elephant and the Flea*, tackles this head-on:

> I am still at a loss to understand why shareholders are given such priority in the Anglo-American version of capitalism. It is not as if they actually "own" the company in any real sense. They haven't in most cases even provided it with money. The first shareholders of each business did indeed give the company money in return for its shares, but thereafter those shares changed in hands through the various stock exchanges without any more money going to the company. *The shareholders aren't financing the business, just betting on it.*[17] (italics added)

One thing seems certain to me. If businesses were owned by the people who did the work, if the people were no longer subjects, the rewards that resulted would be distributed far more equitably than they are today. That could only be a good thing.

In *Democracy at Risk*, author Jeff Gates writes about what a *really* democratic society would be: "Democracy is not a destination; it's our manner of traveling. It's not so much something we *do* as the way we *are*. And the way we are *in relation to others*. Therein lies its sweetness."[18]

It's an apt description of the sweetness of employee ownership that I sense at South Mountain Company. It's the way we have come to be. Employee ownership has become a part of our identity, as individuals and as a company. The searching-for-democracy journey we've taken has stimulated us to articulate and understand our common purpose as we try to find fairness, transparency, and shared responsibility. It has brought a sense of completeness to the company.

Together we've become, at once, better problem solvers and better dreamers. There's a lot to be said for ownership and the responsibility it encourages. As someone once observed, "In the history of mankind, nobody has ever washed a rented car."

Take 'er slow and easy. (Photo by author.)

⌐3⌐

CHALLENGING THE GOSPEL OF GROWTH

Years ago we were designing a house for new clients. The process was going poorly. Our clients wanted to build at a beautiful spot on top of a hill. We proposed to site the house beside the hilltop, so that the lovely area on top, capped with a huge glacial rock formation with a view, would be preserved. They did not share our perspective. They could not believe, even after we presented convincing photographic evidence, that there was a design solution that would, at once, preserve the cherished hilltop landscape and secure the view they desired. I wondered whether we should end the engagement. Given such a fundamental design disagreement and lack of trust so early in the process, it was doubtful the process would go well. On the other hand, this was a big project, and we were counting on it to provide a significant chunk of our workload for the following year to keep our growing workforce busy.

I brought my partners to the site. We sat on the big rock and considered the problem. They shared my view that our design solution combined responsible use of a beautiful site and sensitivity to our clients' needs. We understood that if we withdrew from the project at such a late date, we might not be able to replace the work quickly enough and might run short of work sometime the next year.

We mused a bit. The silence was broken by my oldest partner, who speaks bluntly.

"Let's shitcan it," he said.

The next day I met with our clients and said, "You know, this isn't working the way we anticipated. Before we dig the hole deeper, let's just call it quits." They were surprised, but after some discussion we agreed that it would be better to part company.

As it turned out, we were lucky, and another opportunity quickly filled the gap. We learned to trust our intuition when it told us not to risk the quality of our work in favor of security and growth.

Until that time we had responded directly to demand. When work was offered, we accepted it, and when the volume of work required expanded capacity, we grew. This was standard operating procedure and we had no reason to question it. It was thrilling to have the opportunity. But this incident helped us contemplate the effects of growth, and we began to wonder whether this passive approach made sense for us. We began to examine growth rigorously and evaluate the benefits and detriments.

It may seem odd for a company with thirty employees to have a self-conscious concern about growth. Maybe it's why we've remained so small. While the potential to expand has been steady, we have scrutinized it carefully.

I do not know, from experience, what it would be like if our company were several times—or many times—larger than it is, so it's hard to talk with certainty about the value of smallness. But I have suspicions. I suspect that we could not retain many of the qualities we value if we were significantly larger. But it is a cherished business doctrine that without growth, a company will perish. Many ecologists and a few intrepid economists question whether the planet can sustain a global economy that enjoys perpetual growth, but the idea of individual enterprise growth is rarely challenged in the world of business. I have searched business literature and found surprisingly little that questions the advantages of growth, or that considers optimization of size. In fact, conventional wisdom implies that small businesses are those that just haven't had greater success yet.

Not that we don't favor some kinds of expansion—we do. But we do not embrace unrestrained growth for its own sake (Abbey's "ideology of the cancer cell"). We grow to achieve specific goals, but we are aware that when we choose to increase in size, we may disrupt and endanger treasured qualities. Such concerns do not imply that we must limit development. Economist Herman Daly makes the distinction by explaining that to grow means to increase in size by the assimilation or accretion of materials, while to develop means to expand or realize the potentialities of; to bring to a fuller, greater, or better state. Our planet, he explains, develops over time without growing, while our economy, a subsystem of the finite and nongrowing earth, must eventually adapt to a similar pattern.[1]

If we apply Daly's insight to our companies and look at the implications of growth and the possibilities for development without expansion, we might conclude that remaining small, manageable, and familial has concrete value.

One of the few proponents I have found for limiting business growth is Jamie Walters, the author of a book called *Big Vision, Small Business*. She compares the concept to precious jewels: "It's more a matter of polishing a gem and perfecting its facets, if you will, than of acquiring an ever-expanding number of gems regardless of quality or despite the fact that they might be permanently depleting the mine."[2]

The apparent lack of questioning about the nature and benefits of business growth, however, may simply indicate that the literature lags behind a changing conventional wisdom. In the lead article in a recent issue of *Inc.* magazine titled "America's Favorite Hometown Businesses," the magazine's editor-in-chief, George Gendron, says:

> Wherever I go these days I run into founders who say that getting big fast is not a part of their business plan. They care about financial performance, but they're equally devoted to building a company that promotes personal and professional development, that fosters close relationships with their community, and that gives them pride and satisfaction they haven't been able to find elsewhere. . . . What they lack is business legitimacy. There's absolutely no reinforcement for such thinking in the mainstream culture, and precious few role models for founders who choose such a path.[3]

There is intense debate within the movement for socially responsible business about a parallel growth-related issue: how to keep control of socially responsible businesses as they grow, and how to keep their original values intact. Scale is a critical issue. Many companies that start off with a mission and find early success feel that they must go public to finance expansion. Once they do, they are vulnerable to buyouts by larger companies and subject to corporate law that requires a publicly held company to prioritize profits for shareholders. The takeover of Ben and Jerry's by

Unilever is the most well-known example, but there are countless others. Many small natural and organic food companies, like Stonyfield Farm, Odwalla, and Cascadian Farm—which have been emblematic of independent, live-your-beliefs-no-matter-the-consequences commerce—are now owned by the likes of Coca-Cola, Groupe Danone, and General Mills. The extent to which their freedom to embed their values in their company and their brand may be compromised by their growth is a question.

Faced with such issues, some companies have taken a different approach. Seventh Generation, the Vermont purveyor of environmentally friendly household products, went public in 1993 but saw where that path was leading and was in a position six years later to begin to buy back its stock. The company returned to private ownership and is now charting its own destiny. Patagonia, a pathbreaking environmentally and socially responsible company, has always been privately and very closely held, so when they decided to make a costly shift to organic cotton to satisfy their mission, they were free to take the plunge.

There are no outside investors and no nonemployee board members at South Mountain. Each owner is an employee. We decide what kind of business ours will be. The decisions are partly economic and partly philosophical, and the people making them have well-aligned interests. Our considerations have led us to believe that if our business practice is not governed by an unquestioned growth imperative, we will have greater flexibility and freedom and the character of the business will better match our aspirations.

I am not suggesting that every workplace should be modest in scale. An unquestioning attachment to smallness seems as careless as an equivalent affinity for unconsidered expansion. In our case we believe that excessive growth may narrow our horizons and limit good things like invention, personal fulfillment, and the overall quality of our workplace and our products. Most people I talk to want these good things in their work but find it hard to resist the tug of other forces more persistent. Too often we tend to grow for increased profits rather than to stabilize and improve proficiency. I am profoundly grateful to have partners who are committed to helping one another resist those forces, in favor of a different direction with other rewards.

Why Grow?

Staying small is about realizing when we have enough: enough profits to retain and share, enough compensation for all, enough health care, enough time to give our work the attention it deserves, enough communication, enough to manage, enough headaches, enough screwups. In *The Hungry Spirit*, British business philosopher Charles Handy says:

> In most of life we can recognize "enough." We know when we have had enough to eat, when the heating or air conditioning is enough, when we have had enough sleep or done enough preparation. More than enough is then unnecessary, and can even be counterproductive. . . . Those who do not know what enough is . . . do not explore new worlds, they do not learn, they grow only in one dimension.[4]

Sometimes frantic growth, I think, becomes a purpose in itself, or the perversion of other purpose. For example, our purpose might be to make the finest bagel or supply the best mortgage. But why do we need to produce *all* of either? Why not make just *enough*? The wish to make the best of a product and the wish to make all of a product may each preclude the possibility of the other. It may be impossible to satisfy all the demand for your excellent product without compromising essential elements of product quality. A different approach would be to learn how to do it, share the learning with others, and thereby encourage the establishment of small bakeries and banks embedded in their locale, well positioned to make the best bagels and mortgages for the people they serve.

Some say that to argue about growth in commerce is spurious. Of course you have to grow, they say: "Nature demands growth just as business does." I say, "That's debatable." Wall Street demands growth; business does not. Neither does nature. Nature seeks optimized growth and imposes limits. In the book *Upsizing*, author Gunter Pauli points out that if an oak tree grows to 150 feet, it is strong enough to resist wind, wear, and tear. But it doesn't grow to 1,500 feet, even when nature provides sufficient nutrients. Instead, it provides room for ten other trees. If it grew to 1,500 feet, it would become too fragile and lose its resilience and stability.[5]

Nature has many inherent limits that identify optimal size for different organisms, and we may be better off if we do the same in our organizations and businesses. As business ecologist Paul Hawken once remarked, "Do you want to be a mushroom or an oak tree? Spores beat out acorns every time in growth rates, but never in longevity or durability."[6]

Why do most businesses want to grow? Sometimes there are legitimate reasons that make it necessary in order for a business to survive. Chroma Technology Corp., an employee-owned company in Vermont that manufactures and supplies specialized optical filters for microscopes, must respond to the industry it serves. As the microscope manufacturers grow, they demand more filters. If Chroma can't supply them, they will lose their accounts. Their position in the supply chain requires growth.

The Weaver Street Market, located in suburban Washington, D.C., had no intention of expanding, but a large development that combined residential, commercial, and retail uses was completed nearby and its residents wanted a market. They tried to get a major chain to open a store in their area, but none was interested. So the neighborhood asked Weaver Street to open a second market, and six hundred subscribers signed up to finance the start-up. The residents of the community put their money where their mouth was. How could Weaver Street refuse to offer the service?

More often, however, it seems that the pursuit of happiness has become, for many, synonymous with the accumulation of wealth and power. Maybe it's just because we've been led to believe that we're supposed to grow, supposed to win in the competition of the survival of the fittest.

Our inquiry need not be about growth versus no growth; it better serves us to think about the quality of growth. Some things we want to grow and some we do not. We want to increase our responsiveness, our satisfaction, our effectiveness, our reputation, our legacy, our sense of accomplishment, our relevance, our capacity to improve the quality of our products, and our contributions to good lives for our employees and our community. We do not want to increase our waste, our pollution, our unfulfilled commitments, our stress levels, or our callbacks.

Charles Handy thinks broadly about expansion. He believes that growth can mean not more of the same but "leaner or deeper," supporting improvement rather than expansion. Bigness, he maintains, can lead to reduced focus, excessive complexity, and less effective control. He goes on to say:

> Once big enough [businesses] can grow better, not bigger. It is a formula which Germany's *mittelstander* (small family firms) have tried and tested to great advantage, content to corner and dominate one small niche market, through constant improvement and innovation. Rich enough, and big enough, they concentrate on the pursuit of excellence, for its own sake as much as anything.[7]

Handy's assessment is consistent with Daly's distinction between development and growth. Opportunities for development without growth are legion.

South Mountain and Growth

The growth question came to a head for us in 1994, during a tumultuous period for the company. We had taken on several large projects that had caused us to double our revenues and add employees. Four new owners were coming on board. There were dissatisfactions and unusual stresses. People were feeling that "things are different than they used to be," and there were negative reactions toward new formalities and management systems. There appeared to be a general sense that we had grown too much, too fast.

At a company meeting, we hung a sheet of paper on the wall with a heavy horizontal line and an arrow at each end. The left end said, "Decrease size to 1990 level." The right end said, "Continue slow growth." A vertical line marked the middle, which said, "Maintain present size." Each person was given a sticky dot to place somewhere along that continuum. After all were done, we stood back to see the results, and we found that most of the dots were collected just to the right of the vertical line, a few were scattered at locations farther to the right, and two were right on the "maintain" line. Nobody had placed a dot to the left of the line. This was interesting. It was different from what I thought I had heard.

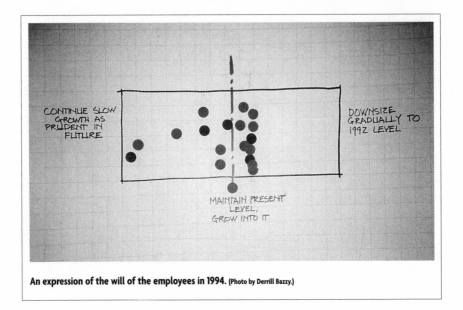

An expression of the will of the employees in 1994. (Photo by Derrill Bazzy.)

The graphic expression of the group's desire was right in front of us, and the discussion that followed clarified what it meant. The general group will was that we should back off on the accelerator a little, adjust our-selves to the recent growth and change, err toward caution, and slow down a bit. Those who leaned toward more growth had specific reasons: with size, they reasoned, we had more ability to do community work and the potential for greater profit sharing.

Since then, we have held a similar meeting every few years. The kind of growth we want to carefully regulate is specific. If we can increase revenue and/or profits without increasing the number of employees or the diffi-culty of working conditions for existing employees, we consider it positive growth or, I should say, development. Adding employees always means greater management burden, and adding workload without increasing staff means added stress. When we hire an employee, we are extending a commitment to that individual. We are not expecting to let him or her go sometime in the future. In thirty years, we have never had to lay off anyone due to lack of work. I don't necessarily think we can keep this record intact, but we'd like to. Fluctuations in workloads and new opportunities can be handled in a variety of ways, but not by adding employees unless we are willing to accept permanent (as best we can tell) growth. On the other hand, if we grow our obligations without adding employees, we have to work longer hours, juggle more, be less attentive, and possibly watch our work suffer. Nobody feels good about that. Achieving a delicate balance among number of employees, workload, and workplace harmony is a constant and unremittingly challenging pursuit.

For the past decade we have mostly agreed that growth at a snail's pace is appropriate. We have aimed to be steady but deliberate. Last year, however, when the company met, a new consensus emerged. We agreed that, for now, we have reached an optimal size.

Is this a conservative position to take? Yes. But you can't casually say "Let's shitcan it" if one hundred or more families are dependent on the company for a paycheck. You *can* turn down inappropriate projects when there are only thirty families involved and you have a reserve fund that can help navigate the rough waters of a slowdown. Strong financial position and small size preserve options.

But isn't this position radical as well as conservative? I think so. To forgo opportunities for growth means the employees of this company have chosen to value the quality of their work life over the size of the potential compensation that might come with more growth. They have chosen the team of oxen over the flatbed truck, because in this case the oxen are the appropriate means of getting the job done.

At the same time we try to remain flexible and open to the excitement of new opportunity. We have found that we can increase capacity without growth by developing new kinds of connections. Over the years we have cultivated a group of relationships with small architecture firms and craft-based general contractors. If we wish to accept work that will exceed our capacity temporarily, we ask these companies to join us. They are microbusinesses with one to four people. We try to make it attractive for them to join our office staff or job-site crews. For them it's an opportunity to work with a larger company without losing their independence. They like the camaraderie and new learning opportunities. We are able to expand and contract without disruption or permanent commitments, and we benefit from exposure to the ideas and methods they bring. These are not temporary employees; they are independent practitioners with whom we create long-term relationships. They're like visiting professors, and we all enjoy the exchange.

Each time we determine that it makes sense to grow, it's with the understanding that people who come to work at South Mountain tend to stay. Before hiring, we must ask, "Are we hiring this person because he or she solves an immediate problem or need, or are we hiring because this is the kind of person with whom we'd eventually wish to share ownership?"

Our decisions about growth do not spring from a firm analytical basis or an identifiable straight-line logic. We are engaged in the exhilarating work of figuring out what we want to be, what we are able to maintain, and how we can get the most—every day—from the time we spend at work. This requires people who are unencumbered by rigid doctrine and empowered by the responsibility for their own workplace satisfaction. It has become a more appealing task to try to profit within the confines of small size than to profit by growing; it feels more humane, more manageable, and more challenging than pursuing maximum profits. Even at our current small size, we are divided into crews and departments—several site crews, the shop group, the office, and the design group. If we were much larger we would have to break into more small groups and become more fractured, and communication among the parts would become harder. We might be willing to suffer these difficulties if we sensed compelling advantages, but at the moment we can't see what these compelling advantages might be.

Making Opportunity

When we talk about emphasizing quality over quantity, there is a parallel between the houses we make and the business we have become. Gendron, in his *Inc.* magazine article, says of our field of residential design and building, "I've come to believe that trends in residential architecture are harbingers of cultural change. While the size of the American family shrank by one third over the past four decades, the size of the average new house in this country grew by 50%. Lately, however, the tide has begun to turn. Books about cottage homes and little houses are selling briskly."[8]

The astonishing and unforeseen success of architect Sarah Susanka's

The Not So Big House and her several follow-up volumes emphasizing quality over size is testimony to this trend. The dynamic craft of creating high-quality compact houses (or anything else, for that matter) is deeply connected to the idea of making businesses that also prize quality more than size. Susanka says, "It's time for a different kind of house. . . . A house for the future that embraces a few well-worn concepts from the past. A house that expresses our values and our personalities. It's time for the Not So Big House."[9]

In our work we embrace the same standards she proposes: small scale, high quality, protective land use, environmental care. Why wouldn't we want our business to have the same attributes? There's a place for the not-so-big business as well. This is not to say that there's no place for good big businesses or good big houses, but only that sometimes they are unnecessarily so, and often the goodness declines with the bigness.

With small scale come particular challenges in providing opportunities for career development. We need to do some of those things that large companies sometimes do well, like offering training, providing varieties of work experience and room to move within the company, and contributing generously to community projects. We can do these things only modestly.[10] When we manage to provide diversity of experience and opportunity for movement, the talents and interests of our employees determine these directions.

Ten years ago, when our bookkeeper left, we hired Deirdre Bohan to replace her. Within a year she had made what had been a forty-hour job into a twenty- to twenty-five-hour job. She came to me and said, "I don't have enough to do." I asked her what she wanted to do. "That's up to you," she said.

"No," I said. "I mean what do you *really* want to do?"

She told me she had an interest in practicing interior design. We had been wishing to add an interiors component to our work and had always done it informally and unsystematically, but we had never before had the impetus to pursue it seriously. We decided to devote the time Deirdre had created—twenty hours a week—to her education. She put together a well-rounded program that used the resources of several design schools. She pursued it vigorously and she now runs a thriving interiors department. We do the complete interior on virtually every project we undertake, enhancing our impact on the final product while also creating a new profit center.

We sometimes assign unconventional duties based on particular capabilities. Michael Drezner, a carpenter and owner, manages the investment of our pension fund and reserve fund. He has been a long-time student of finance and the stock market, and over the years he has developed his skills by managing these two funds in collaboration with a socially responsible investment advisor. The two accounts contain approximately $1,500,000 at this point, so this is no small burden to bear.

While these two examples demonstrate the possibilities, they also highlight the danger of staying small and hiring from within. What if the efficient bookkeeper is a mediocre interior designer? What if the steady carpenter invests money unwisely? So far, our experience has been positive whenever we've made such moves. Recently, we have made education a priority, so that we can create more opportunities for employee enrichment. If we continue to hire for ownership potential rather than specific skills, we are likely to have a versatile workforce in years to come.

There are still times when I wish we were a larger company with the ability to offer a broader range of possibilities to our people. For example, we had an employee who was hired as our receptionist. He was an engaging thirtyish "Gen Xer," social, outgoing, funny. He answered the phone, directed traffic, planned our parties and events, and did general office tasks. He managed certain kinds of communication within the company. But he had little aptitude for or interest in bookkeeping and development of office systems, the areas we needed him to grow into, so there just wasn't a whole job for him. I thought, "If we were a larger company, one with perhaps a hundred employees, we could keep him, and his job description could be 'receptionist, event planner, and generally cool guy to have around the office.'" We are always learning to deal with the limitations that come with our small size. In this case, the employee left the company and has enrolled in architecture school, so maybe it's all for the best—he was inspired by his time here to find a new path to fulfilling work.

Rule of 150[11]

Growth can be an extreme sport. When a company is growing quickly there's a thrill a minute. It's the same type of sensation many people seek by climbing a mountain or soaring off a cliff clinging to a hang glider. Some of us are willing to forgo such thrills in our work in exchange for familiarity and stability. Some try to get the best of both, and these people have made important discoveries.

When organizations become large, there is often the concurrent inclination to make small units within the larger structure to maintain qualities like conviviality, effective communication, and flexibility. Malcolm Gladwell's *The Tipping Point* explores how little changes can have big effects and turn ideas, products, messages, and behaviors into major trends. In the book Gladwell writes about the theories of anthropologist Robin Dunbar, who, in the interest of learning about optimal size, has studied how groups of varying numbers work. A striking collection of examples supports his conclusion that there is a Rule of 150, which says that 150 is the maximum number of people who can share a social relationship with each other. Therefore, organizations work best if they remain within that rough limit.

The number reveals itself in a variety of interesting settings. Dunbar looked at twenty-one different hunter-gatherer societies around the world and found that the average number of people in each village was right around 150. The pattern holds true for military organizations, whose planners have a rule of thumb for the size of a functional fighting unit: 150 to 200 soldiers. Reduced hierarchy, fewer rules, and fewer formalities are required for the group to function as a team if it remains at that size. Group behavior operates on the basis of personal loyalties and relationships in a way that is impossible with larger groups.

The Hutterites, a religious group that has lived in self-sufficient agricultural colonies in Europe for centuries and in North America since the early 1900s, also adhere to this concept. When a colony begins to approach 150 members, their policy is for that group to split and a new one to branch off. A leader of a Hutterite colony told Gladwell,

> Keeping things under 150 just seems to be the best and most efficient way to manage a group of people. When things get larger than

that, people become strangers to one another . . . you don't have
enough work in common, and . . . that close-knit fellowship starts to
get lost. . . . You get two or three groups within the larger group.[12]

The W. L. Gore Company stands out as the most visibly committed
proponent of this approach in business. Gore, a privately held company
that is responsible for Gore-Tex and a variety of high-tech products, is
widely considered to be one of the best-managed American companies
and one of the best companies to work for. It is a billion-dollar enterprise
with seven thousand employees in forty-five locations. Each company
plant has no more than—you guessed it—150 people. The company's
management feels that it has been able to retain the feeling of a small
company by adhering to the Rule of 150 and by spreading ownership and
responsibility throughout the company. The employees of each plant are
called "associates." They work together with little hierarchy and widespread involvement in decision making. The company is owned by the
Gore family and the associates. The Associate Stock Ownership Plan
(ASOP) provides equity ownership and financial security for retirement
for employees. All associates have an opportunity to participate in it. The
stock is privately held; it is not traded on public markets. The ASOP, not
the Gore family, is the majority owner of the company. This is an unusual
company in many ways, and it has neatly balanced an absolute limitation
on unit growth with an open-ended approach to overall company size.

Other businesses have also found the Rule of 150 to be useful.
Chatsworth Products Inc., a medium-size, highly successful California
manufacturer, is 100 percent owned by its employees. It is also broken
into smaller units. In *The Soul of Capitalism*, William Greider tells of
Chatsworth's successes and quotes CEO Joe Cabral:

> Most business managers think I'm crazy. I've been told, "Joe, you
> can't do these operating units as small as you envision them." Our
> philosophy is, we don't want any unit beyond a certain size—our
> magic number is around 150 people—because you need an envi-
> ronment of family. Everyone working together knows each other,
> they care about each other, and they're willing to help each other

out. When you get above a certain size you end up with walls and workers become just faces rather than people or[,] god forbid, numbers.[13]

For some businesses, the W. L. Gore approach and the Chatsworth method may be fine models for growth. For others, smaller scale is absolutely essential to the character of the business. An extreme example is Phil Goldsmith's medical practice in Boston. For the past decade or so, my wife and I have had an annual appointment with him for a full physical exam. A visit to him is different from a visit to most doctors. He talks to you, looks you over, pokes at you, and hunts for hints, signs, and clues. He's an investigator and a healer. He does everything himself, without nurses or assistants, so he gets more time with you to talk and probe and discover. Yet this is no country bumpkin but, rather, a highly regarded member of the Boston medical establishment. Phil says, "My practice isn't for everyone. Some come for the first visit and just can't wait to get the hell out of here, and never return." It's because he's demanding. For example, he says, "I don't allow colon cancer in my practice. It's entirely preventable." When you turn fifty you get a colonoscopy, dammit, and every five years thereafter another one.

During our most recent visit I talked with him about my view that some businesses make more sense on a smaller scale. He said, "It's just like my work. I do what I love and I'm good at it. If I had a large practice, I wouldn't be doing the same thing anymore." Phil Goldsmith's practice is highly competent, carefully networked (he sends his patients to a variety of handpicked specialists as necessary), information rich, and personal. In some ways, it's almost an anachronism, but it shows that you don't have to be big to get good results, a devoted following, and even prestige. In fact, ironically, ignoring the pursuit of status sometimes brings it. It's true in his case. Many doctors with whom I've spoken are upset and disheartened about the bureaucratization of medicine and the loss of control imposed by HMOs and insurance firms. Not Phil. He's upbeat. He has no complaints because he has made the tough decisions—and at times the sacrifices—that keep his practice successful and satisfying. And free of colon cancer.

Franchising Democracy and Local Wisdom

In this chapter I have argued that it is often unwise and counterproductive for small businesses to grow. There may be other ways the argument falls short, but here is one for sure: If we develop a service, product, or business model that improves people's lives, we would want to make it available to as many people as possible, right? But if we grow our businesses to do so, they may become distanced from place and people and lose qualities we wish to maintain. I wonder whether we might be able to keep our cornerstone business principles intact, while expanding influence and effectiveness, by using the practice of franchising in unconventional ways.

Franchising is familiar to all of us. It's a popular and successful form of business. More than 35 percent of all U.S. retail sales are made by franchises, and the practice continues to grow. It became prominent in the 1950s, but by the early part of that century it had already dominated key sectors of the economy, like automobiles, gasoline, farm machinery, and sewing machines. By the end of the '60s it was used to sell nearly every kind of goods and service. Franchising combines large and small businesses into a single administrative conglomerate. It's a way to grow businesses quickly by creating many independent enterprises that must, however, conform to rigorous standards of operation designed and enforced by the parent company. The term originates from an Old French word meaning "to make or set free," but, ironically, most franchises have little freedom.

A franchise is a business that is authorized to sell another company's goods or services in a specific area in exchange for an agreed-upon start-up fee and regular royalty payments. A company that has been successful with a product or service can use this method to extend its business geographically. There are two kinds of franchises: those that distribute products and those that distribute business formats. Ford is a product franchise; McDonald's is a business format franchise. The product concessions came first, but as company owners learned that the franchise itself could be marketed as a distinct product, the business format led to a wave of new franchisers—hair salons, sewer cleaners, insurance firms, tax preparers, pest control companies, copy shops, pizza shops, photo processors, Laundromats, and tai chi instruction.

Franchising offers a low-risk avenue to business ownership. In one sense it has had a democratizing effect, since someone with little to invest can go into business and have access to some of the benefits of highly capitalized business operations. It is also a way for small business to adapt to competitive conditions. At the same time, franchising contributes to the rapid homogenization of the nation's (and now the world's) business landscape. Consuming has become an experience that is repeatable no matter where you are—the same signs, same advertising, same products, same experience. Our predisposition toward branded goods of uniform and proven quality over unique and unfamiliar products makes franchising a logical way to build businesses.

The character of consumerism may be changing, however. There is evidence that people are now beginning to want differentiation and customization and are searching for authenticity and choices. In her book *The Substance of Style*, social commentator Virginia Postrel says,

> The increase in aesthetic pluralism spurs competition to offer increasing variety. . . . The holy grail of product designers [has become] mass customization. Industrial design guru Hartmut Esslinger . . . imagines modularly designed products that could be recombined "to offer 100,000 individual versions," expressing as many personal styles. "Mass production offered millions of one thing to everybody," writes another design expert, upping the estimate. "Mass customization offers millions of different models to one guy."[14]

If Postrel has correctly identified a trend, franchisers may have to change. It may not always work to offer the monotony of sameness. But for now, the conventional franchising wisdom has a number of key tenets: consistency among outlets, identical customer experiences at all locations, and system-wide decisions made by the central office. Don't become a franchisee if you are an entrepreneurial type who likes to make your own decisions.

As a franchisee, you may even be told when to take out the garbage and what to do with it. Dave Thomas, the late founder and CEO of Wendy's, whom we saw in television ads standing at the grill, said, "Training our people is the most important thing we do. And the training never ends. We even teach them how to clean the tables. Believe it or not, there's a right way and a wrong way."[15]

Is there?

Great Harvest

The antithesis of the Wendy's way can be found in the philosophy of an exceptionally intelligent company called the Great Harvest Bread Company. This small bread maker, headquartered in Dillon, Montana, has developed a unique method of franchising and communication that is as different from that of Wendy's as fresh-baked whole-wheat bread is from burgers and fries. In its approach to franchising there are elements of Postrel's ideas of mass customization. There are 186 Great Harvest bakeries across the country selling some of the best bread you will ever taste. Each is a franchise, and more are opening on a regular basis. The owners of Great Harvest believe in their product, but they also believe in freedom, innovation, and the integrity of the community of owners who are their franchisees. All the bakeries are locally owned and operated and are unconstrained except by several basic principles. Great Harvest has a simple franchising agreement that derives from a foundation of respect and trust. The only required provisions are to ensure the quality of the product: you must buy your wheat from approved sources and you must grind it fresh every day. That's it, and further, there is a provision in the contract that says, "Anything not expressly prohibited by the language of this agreement is allowed."[16] Have you ever heard of a contract like that?

You will see that I've oversimplified somewhat, but it is entirely true that Great Harvest is franchising both a distinguished product and a progressive business ethic in a fundamentally unique way. I first encountered the company in the book *Bread and Butter*, the story of Great Harvest told by Tom McMakin, who started working there in 1993 as the company's newsletter editor and ended up managing the business for its founders and original owners, Pete and Laura Wakeman. I made a point of visiting a few Great Harvest bakeries when my daughter and I were looking at colleges out west. Each was different, each was a great place to visit, and the bread was uniformly superb. At one point, between cities, I asked Sophie whether she was hungry.

"Starved," she said.

"Wanna stop and eat?" I asked.

"No, that's all right. I'll just butter us up another slice." Great Harvest. We lived on it for days.

I wanted to know more. I spoke to the current CEO, Mike Ferretti. He and a small group of investors bought the company when health reasons caused the Wakemans to sell. Mike is careful and thoughtful when he speaks about Great Harvest, but his modesty does not mask his pride. He thinks of McDonald's, Starbucks, and other similar chains as being in one category and Great Harvest in another. The differences are striking. He, like McMakin, thinks of Great Harvest bakeries as "freedom franchises." In every way the company encourages independence and initiative. Other franchisers have regimented corporate regulations and rigorous advertising rules. Franchisees must toe a well-defined line. Company field agents are cops whose job is to patrol and enforce.

At Great Harvest, quality begins with the selection of franchisees. Mike says they "take the selection process to an extreme." They feel an ethical obligation to do so, for both the company's sake and the sake of the franchisee. To open a new Great Harvest bakery is a major investment and life change, and the central office wants to be as certain as possible that success will follow. There is a series of interviews and visits back and forth. The interviews "are two-sided; we have to be right for each other." The new association is treated as the beginning of a partnership, just as hiring new employees at South Mountain is the process of identifying future owners. Once the selection is made and the location is approved by the parent company, there is extensive training and support, which continues during the first year after the new bakery opens.

A recent development at Great Harvest is the implementation of quality standards for both customer service and products. These standards have been developed in consultation with the outlets and are implemented and monitored by "a council of bakery owners," not by the front office. "They're the ones who can best evaluate what's most important. It's a jury of peers," says Mike. So after bakeries do the required—source the right wheat and grind it fresh daily—they are expected to produce superior bread coupled with excellent service, but each decides how to achieve it. There are no standards for recipes, store design, or product selection. The philosophy is "Do it your way, but do it well." The central office establishes "handrail" philosophical guidelines, not strict rules. There is plenty of information and assistance available if you want it.

Another critical part of the Great Harvest system is a self-regulating conversation among owners that has been going on for many years. All are connected to a company-run computer network over which, via e-mail, ideas, problems, and issues are shared and discussed. This still-maturing system has become the primary means of communication within the company. Everyone hears and knows as much as they wish to. The front office has never removed a single comment, although, Mike says, "sometimes we've sure wanted to." With this approach, the field agents have a different role than policing. As McMakin describes it, "What our field representatives do on their visits . . . is different. Their aim is not to assess how well an owner is complying with the corporate manual, but to act as bees, buzzing from bloom to bloom, cross-pollinating as they go, making sure knowledge passes from bakery to bakery."[17]

The franchisers' job is to create opportunities for owners to exchange ideas. In addition to the online discussion, they organize conventions, regional meetings, and owner-led trainings; facilitate group buying and marketing cooperatives; and pay for owners to visit each other's stores. The result of all this, says McMakin, is that

> [what] makes Great Harvest hum is that owners of bread stores have the best of both worlds—they are in close association with others who are doing exactly what they are doing, and this proximity sparks in them all sorts of interesting ideas. But they also are completely autonomous, free to implement in their stores the ideas that excite them. Being organized as a freedom franchise allows all of us in Great Harvest to combine *quick learning with the power of rapid adaptation.*[18]

Great Harvest is significantly different from other franchisers in several other ways. The royalties that the bakeries pay *decrease* every five years.

As a "reformed CPA," as he puts it, Mike Ferretti recognizes that this is unusual. Yet it's practical and logical, too. It is only partly a reward for length of service. Long-term franchises need less support, which reduces the front-office workload.

Great Harvest has unusual aspirations about growth. Its long-term goal is to open, on average, two new bakeries a month in perpetuity. No more, no less. That's how the owners feel the company can be most effective. This means, of course, that as each year goes by the company's growth *rate* is actually declining. By operating in this steady manner, the owners do not need to swell their office staff (there are currently only twenty-eight central-office employees with 186 bakeries operating) and face the prospect of having to grapple with market fluctuations by downsizing. The company remains intimate, personal, and nimble. Instead of committing to the growth of their own business, the owners of Great Harvest have committed to the establishment of many small businesses, each unique, that are part of a learning organization.

I asked Mike whether he knew of other companies like Great Harvest. His answer: "I don't want to be glib about this, but I'd say there are probably many, but none, I'm betting, with more than fifty franchises." He said that many companies will do anything they can to attract early franchisees, but once they get into the twenty- to fifty-outlet range and there is established demand for their franchises, the goal of the companies shifts to the exercise of greater and greater control and faster and faster growth. He knows of no other company of its size that operates like Great Harvest. A central aspect of the company's long-term business plan is to remain private, so the pressures of Wall Street and absolute obligations to maximize returns to stockholders do not overwhelm them.

It is valuable to compare the Great Harvest experience with that of the Body Shop, a well-known icon of socially responsible business known for its activism. Advocacy positions covered their trucks and store windows in the early days. Body Shop founder Anita Roddick also had a commitment to independence, irreverence, and individuality for her franchisees. She and her husband were strong on social mission, but when it came to business, they adopted the conventional paradigms of big growth and going public. Once beholden to Wall Street, they ran into problems of a sort that Great Harvest has never encountered, at least not yet.

The idea of franchising South Mountain has been tumbling around in my head ever since architect Gordon Tully came up to me after a talk years ago and said, "You should franchise what you do."

I tossed it off with something like, "We don't even know what we're doing ourselves—how could we franchise it?" But the notion returns whenever we have company discussions about creating new jobs that can serve people for their full careers.

Roughly half our employees are carpenters; some are starting to get too old for this work. Could some of us move from making buildings to helping others make satisfying and effective design/build businesses? Several events caused me to think more seriously about the idea. First I read *Bread and Butter*. I had never before imagined or encountered a conceptual framework for spreading a business practice that made sense to me, but this one did. We don't make a product—like bread—that can be readily franchised, and our business is so decidedly place-based that it does not seem to lend itself to franchising. I'm beginning to wonder, however, whether that's just the point: could the thirty years of information and experience we have accumulated—about making a small, locally committed, democratically organized design/build business that's profitable and long-lived—somehow be sold as a "knowledge franchise"? I can't quite envision it yet.

Small Time

Nor can I envision our company with 150 or more people. I can almost imagine it with fifty, or maybe sixty. Even now I don't always remember the names of all the kids of my workmates. Since many people are scattered at different job sites, I may not see someone for weeks. Occasionally it takes months or years to have follow-up conversations to the mutually probing exchanges we had around the time of a person's hiring. I wish I knew everyone better. I wish I made more time to catch up on people's lives, and shared more of mine. I wish there were more chances to explore the intricacies—the hips and valleys, the copes and scribes, the successes and failures—of the projects they're doing.

The pursuit of concentrated power and wealth may be like chasing a porcupine—if you're not careful, you just might catch it. I've come to believe that there are optimal scales for different businesses and organizations, that we need to think more broadly about the meaning of growth, and that the concept of "enough" has a place in our internal debates. As our ownership pool grows, we may have to expand our ability to create individual equity as the larger numbers dilute the distributions. If one of our goals is to extend our influence through growth, we may have to find inventive new forms of growth, like observing the Rule of 150 or implementing new forms of franchising. Careful examination and control of growth has become South Mountain's second cornerstone, a prominent link in our chain of values. It's a tug on the sleeve that has our full attention; the gospel of unrestrained growth is not the right doctrine for us.

There's a story about a fisherman who was sitting on the beach with his wife one afternoon enjoying the surf and the sun. He had enjoyed a big catch that morning, so he came in for the day. A wealthy businessman heard about his success and approached him.

"Why didn't you keep fishing and bring in twice as much?" he asked.

"Why?" said the fisherman.

"Because you could make more money. Maybe buy another boat and hire some employees."

"Why?" the fisherman asked again.

"You could keep growing, increase profits, and buy more boats. If you worked long and hard at it after some years you'd grow rich."

"Why would I want to do that?"

"Because then you and your wife could retire and relax on the beach," said the businessman.

"But that's what I'm doing now."

You call this work? (Photo by author.)

4

BALANCING MULTIPLE BOTTOM LINES

Not long ago I was deeply moved by a decision of the South Mountain owners.

One of my partners, who had recently adopted a child, presented a proposal for an adoption benefit. It was well researched and he made a good case. Birth parents have all their expenses covered by our medical insurance, while adoptive parents have none. He proposed that our maternity and paternity leave should be formally extended to adoptive parents, that they should receive a $2,500 to $5,000 cash benefit to offset part of their expenses, and that the new benefit should begin with the next adoption. Discussion led to the conclusion that if we're trying to create parity between birth parents and adoptive parents, the $5,000 figure was too low. We settled on a larger number. After a final inquiry to affirm consensus, one of the owners asked, "How many think this should be retroactive?" Nearly everyone's hand went up. Again we probed for consensus and found it. This meant a large unexpected payment would go to the partner who had suggested the benefit. A tear rolled down his cheek, and he sat in silence, unable to speak.

When everyone feels there is enough for them, the impulse to share the abundance has a chance—just a chance, I'm saying—of prevailing. It becomes clear at moments like these that the financial bottom line is only a tool in service of the multiple bottom lines by which our company measures its ultimate success.

Describing the Wine

We all know it at some level, but when we get down to business it is all too easy to forget: using money as the sole measure of prosperity fails to recognize that people have lives, families, and communities. Many people value the quality of their work environment as much as, or more than, the size of their paycheck. In addition to making a reasonably good living, we need to be satisfied by our work. We need to meet the expectations of our clients and business associates, to contribute to the stability and vitality of our community, to care for the environment, and to maintain a workplace that is safe, healthy, and rewarding. We need the pleasure of good service, the joy of humor, the treasure of strong relationships, the fulfillment of collaboration, and the security of stability and longevity. Our enterprise must create sufficient profit to be able to serve all these worthy ends, but profit is simply the engine that drives a bottom line composed of many parts.

Our third cornerstone principle is balancing multiple bottom lines while consigning profit to its appropriate role.

In recent years many businesses and nonprofits and even some governmental agencies have begun to account for multiple bottom lines. Usually they have added environmental and social accounting to the traditional economic tracking to make a triple bottom line. The term *triple bottom line* (TBL) was popularized in 1998 by John Elkington in the book *Cannibals with Forks: The Triple Bottom Line of 21st Century Business.*[1] In the few years since the term has caught on. Some of the larger accounting firms are now "offering services to help firms that want to measure, report, or audit their two additional 'bottom lines,'" according to Wayne Norman and Chris MacDonald, in an article in *Business Ethics Quarterly.*[2] Norman and MacDonald are critical of the way the term has been used, however, since there is no methodology for measuring and comparing social and environmental bottom lines to the accepted accounting standards that define financial reporting. Their research led them to the conclusion that the term *triple bottom line*, as used today, is vague and misleading and promises more than it can deliver. "In short, because of its inherent emptiness and vagueness, the triple bottom line paradigm makes it as easy as possible for a cynical firm to appear to be committed to social responsibility and ecological sustainability."

South Mountain claims no triple bottom line. We are absolutely committed to social equity and ecology as well as profits, but we do not measure and account for our social and environmental impacts, and our relative successes and failures in these nonfinancial pursuits. Rather, I am using the term *multiple bottom lines* broadly and metaphorically. It is an indicator of the variety of parallel concerns, values, and issues that drive us. Each of these is as important as the economic bottom line, but none can exist without healthy financials, which are the only ones we measure. As somebody once observed, "It is easier to count the bottles than to describe the wine."

Description alone, however, is not enough. We can't become who we wish to be just by saying it, and it has taken years of toil and turmoil to find the track that leads us toward being the kind of company we'd like to be. This aspect of building a company is like tending a garden, and each of the bottom lines is a row that needs to be prepared, planted, cultivated, weeded, and harvested. Because we're the owners, this is kind of a kitchen garden. We're doing it not to make goods for sale but to create balance and quality in our own lives and work.

When people pay their South Mountain members' fee and sign the ownership documents, they become legal owners. But they are owners only in a sense. It's like walking out of the attorney's office just after closing on a new piece of property. You may own it, but it's not truly yours until you've experienced a sunset, a full moon, a downpour, a swarm of mosquitoes, a gusty day in the fall, an ice storm. It's not yours until you've come to know the trees and shrubs. It's not yours until you've tended the garden. The property is not yours, in fact, until you have confronted at least one skunk on the path to the front door.[3] That's how it is with our business. Work experiences and shared decisions accumulate within each individual, gradually making the company his or hers. The skunk on the pathway heightens our senses as we tiptoe down the path to ownership. Tending this garden, and taking responsibility for stewardship of a diverse set of commitments, makes multiple bottom lines meaningful and real.

Growing Pains

A February 1993 company meeting, with all employees present, may have been a pivot point in our understanding of multiple bottom lines. That day, grumbling about wages and other dissatisfactions coalesced into productive dialogue. In those days, our company meetings were sporadic affairs. To me, they sometimes felt like being at a dance with no music—we kind of wandered around without purpose. This meeting felt particularly disharmonious. It was suddenly galvanized when Mike posed a series of questions to everyone.

"What's important to you?" he asked. "Do you need greater compensation? Or are other parts of what the company does more important to you? Obviously everyone can use more money. That's not the question. The issue is one of priorities and balance. What are your priorities?"

These were the right questions at the right time. Heads rose, eyes lost their glaze, yawns disappeared, and people spoke from the heart. The few loud voices we had been hearing did not represent the general will and feeling. Among the statements that found widespread support were:

- *We would be less happy at work if South Mountain were less concerned with social progress.* Affordable housing, sliding-scale work, care for the environment, contributions to social causes—all of these were apparently critical to peoples' satisfaction with their work.
- *We should have a formal annual evaluation system and an equitable, accessible wage schedule.* Company members said they wished to see transparency throughout the company, not just among the owners. They also questioned whether our compensation was uniformly fair, and whether our mix of wages and benefits was appropriate.
- *Job security is essential.* Profits were seen by most not primarily as an avenue to their own personal financial independence, but as a vehicle for ensuring the longevity of the enterprise and their role in it.
- *We should continue to build our reputation as a socially engaged company.* It was pointed out that the company's good reputation

was partly a result of our community work; therefore, social activism directly affected job security and profitability.

• *We need a diversity of work opportunities.* People expect to be here for the long haul, and we wanted to help the company create new kinds of work for us to do as we become too old to spend our days running around on rooftops.[4]

The many points of agreement that emerged from discontent at this watershed meeting allowed the board to move forward with a better understanding of multiple needs, and it inspired more company meetings, more engagement, and a new respect—among nonowners—for what it meant to be an owner. I wrote a memo to all employees after the meeting that said, in part (I've paraphrased some of this for clarity and brevity):

> Thanks to all of you for your participation in the company meeting last week. Thanks for your forthrightness, sincerity, heartfelt concerns, and ideas. . . . I believe we've reached a turning point. For a while we've been successfully engaging employees in steering the course of the business when they become owners. We've been less successful at bringing the rest of you into this process. There's a simple reason why it is valuable to do so—it's an opportunity for each of you to assist in making things more like you wish them to be. We'll never build a perfect house, nor will we ever be a perfect business. We'll never even agree about what a perfect business would be (or a perfect house, for that matter). But the expression of your needs, desires, and ideas for improvement will make us a stronger and more responsive company and will make personal satisfaction for each of us more likely.
>
> The meeting pointed out many things that need work. The necessary changes will happen slowly, over a period of years. I'm sure at times it will seem like we're moving backward instead of forward, but you don't discover new lands without losing sight of the shore for a while. I look forward to the journey. Thanks for your company.

At another meeting three months later, we concentrated on compensation and evaluation issues and agreed that competitive evaluations should not be the criteria for pay scales. Specific additional benefits were deemed to be desirable. We also agreed that it was time to put our heads together to define and articulate our mission, our core values, and our purpose. This was a moment of different consciousness. Before that, the questions we had tackled together were mostly about how to do what we do better than before. Now we were asking ourselves, "Who are we and why?" This was key to our unfolding understanding of multiple bottom lines.

The next company meeting, in February (when the Vineyard is always bleakest), evoked an outpouring of protest and complaints. The venting was prompted by the observation of an employee who had begun work with us when he and his wife had taken a sabbatical from their regular life in Connecticut. He was a valuable addition, as he'd been a contractor himself for twenty-five years and had a wealth of experience to share. They had had such a good time on the Vineyard that they went home at the end of the year, finished their nearly restored house, sold it, and moved back here, and he'd come back to work at South Mountain.

Now he said, "Things are different than they were when I was here two years ago." And it's true that they were. He had returned to a larger and busier company, a company that was experiencing new stresses and tensions.

As the meeting evolved, it became clear that the stress experienced by members of the company was coming not only from work issues but also from their lives outside of work. There were new babies, new houses, and additions to old houses. The new employee to whom it felt "different" was a good example. Two years before he and his family were experiencing an idyllic sabbatical in a new and welcoming place. Now they'd uprooted their lives and they were busy buying land, dealing with banks, drawing plans, adjusting to new jobs—nearly everything in their lives had changed. It wasn't only the company that was different.

We learned a lot from that meeting. We found, for example, that our new job schedules were causing trouble. We had never had such detailed schedules before, and everyone seemed to be worried about them. We became aware, too, that people were worried about budgets, which had become more accessible in accordance with the group's wishes. The only thing that had changed was the exposure; putting the schedules and budgets on the radar screen made them available to review and, so, to worry about. What was happening—with overload, schedules, budgets, and wages—was that *information* was beginning to bounce around and reverberate. It had safely rumbled around in my head for years. Then it had gradually spread to the group of owners. Now, as they learned to balance the complexities, the owners were beginning to spread information throughout the company. New knowledge is sometimes tough to swallow. An appetite for information had developed, and despite the new stresses the information caused, or perhaps because of them, the capacity to absorb it began to increase. These tensions reflected our development into what former Royal Dutch/Shell CEO Arie de Geus calls "a learning organization."[5] In our case, the learning was learning to balance multiple bottom lines.

Other emblematic board discussions helped us define our bottom lines. At one point, in 1996, I suggested the use of the phrase *ecological building* in the tagline on our stationery. There was strong negative reaction. A litany of examples was tossed out to point out how distinctly *unecological* our buildings and our work are. We build houses that are sometimes larger than they need to be and use excessive resources. We do a less-than-exemplary job of reusing and recycling construction waste. Everything we do has aspects of negative environmental impact. I responded that our environmental bottom line had received the same degree of emphasis as our social bottom line. There is a strong internal commitment to incremental progress in our use of energy, our approach to waste, and our support for and development of environmentally sensitive and low-impact land-use strategies.

Of course we're not doing a perfect job, but aren't we trying to do the best we can, and doing a better job of it than almost anyone else we know, and therefore don't we have the unassailable right to define ourselves that way?

This tack didn't work; I got nowhere. As I argued my point, I slowly came to realize that the critics were right. It took me a while to get it, but I came to understand their view that it was presumptuous to label ourselves as ecological builders. We should do the stuff, let the work speak for itself, quietly highlight the ecological aspects of our work when the opportunities present themselves, and leave it at that. I admired the passion and precision with which this call for restraint was articulated. It was an important identity issue. As a group, we eschewed greenwashing, preferring silence to inflated or questionable claims, even if we might be the only ones questioning the claims.

I tell this series of stories and events to convey the rhythm of the conversation—which I now am able to identify as being about multiple bottom lines—that has become essential to our company. The dialogue has been the vehicle that carries us to a deepening of our shared values and mutual commitments. We have not made these up; we have discovered them. To track back along the trail that led to them, before I wrote this chapter I spent two long days reading through six large loose-leaf binders of our meeting minutes and attached memos dating from 1985, when we first began to meet formally, to the present. The review was by turns tedious, embarrassing, funny, uplifting, instructive, and gratifying. It's a rich vein, and it was remarkable to mine it all at once. True controversies were rare but significant. There were times of turmoil and times of relative calm. The thread of multiple bottom lines is now recognizable in ways that were not apparent along the way.

Billy and Derrill studying the plans (!) and plotting the next move. (Photo by author.)

Mining Collaboration, Curbing Competition

About ten years ago I read *No Contest: The Case Against Competition*, a book by educator and social theorist Alfie Kohn. This was a memorable experience of affirmation for me. It was as if Alfie Kohn were a new friend, articulating clearly what I felt deeply.

Kohn has been described as the country's leading critic of competition, although he is said to be quick to point out that there is not much competition for the title. In his thoughtful and well-researched critique of our winner–loser society, he shatters myths regularly. Using a combination of his words and mine, I will try to convey to you, in a few pages, what he takes a few hundred to reveal far more completely and competently. I want to do this because Kohn's thesis explains another crucial part of the South Mountain bottom line, and that of other collaborative organizations, businesses, families, and classrooms around the world. Kohn's insights run distinctly against the grain of American culture and question much that we take for granted. They contain valuable information for those interested in a more compassionate economy built around healthier workplaces and communities.

Kohn describes how human nature is characterized more by the urge to cooperate than by the will to compete. He argues that competition systematically damages relationships and crushes self-esteem. Most competitors lose most of the time because by definition not everyone can win. The race to win turns most of us into losers, and as Lily Tomlin once said, "The trouble with the rat race is that even if you win you're still a rat." Kohn puts no stock in the theory that competition builds character. He maintains that competition actually holds us back from doing our best, in school and at work.

My friend Lee points out that, ironically, the act of marshaling the forces of cooperation against the forces of competition might be the ultimate test of character.

If all this is true, or even partly true, why do we hold competition in such high regard? Why has it become, as one author says, "almost our state religion"[6]? Because, says Kohn, when we think about cooperation we tend to associate the concept with fuzzy-minded idealism or, at best, to see it as workable only in a very small number of situations. This association may come from confusing cooperation with altruism.

Cooperation is a shrewd and highly successful strategy, a pragmatic choice that gets things done more effectively, but we have been trained to compete and, more important, to believe in competition. The message that competition is appropriate, desirable, required, and unavoidable is drummed into us from nursery school to graduate school. The results of this teaching, which permeate every lesson, are used to prove competition's inevitability. If you make it so, it will be so.

Kohn recognizes isolated progress toward a collaborative ethic in our classrooms—some teachers are coming to honor cooperation as the primary method for human interaction—but he thinks most teachers don't understand the concept of cooperation. They use it to refer to obedience; to cooperate, for them, is to follow instructions. This misunderstanding means our education often lacks critical learning about cooperative behavior. We are taught how to get the job done and win games, but we are not taught the skills of true cooperation.

I am continually amazed that we are not taught in school how to lead meetings or to be effective meeting participants. Very few people are trained in meeting facilitation. This is an unfortunate oversight because it seems like an essential—and rare—skill. Commerce, government, and community life all suffer from our lack of meeting facilitation and collaborative decision-making skills.

Kohn's findings indicate that success and competition are conceptually distinct and unrelated. He goes so far as to say, "Superior performance not only does not *require* competition; it usually seems to require its absence."[7] He relates the findings of several research projects to support the conclusion, including one exhaustive overview:

> David and Roger Johnson and their colleagues published a[n] . . .
> ambitious meta-analysis (that is, reviews of others' findings) in
> 1981. In what is surely the most conclusive survey of its kind, they
> reviewed 122 studies from 1924 to 1980 . . . including every North
> American study they could find that considered achievement or
> performance data in competitive, cooperative, and/or individual-
> istic structures. The remarkable results: 65 studies found that

cooperation promotes higher achievement than competition, 8 found the reverse, and 36 found no statistically significant difference. . . . The superiority of cooperation held for all subject areas and all age groups.[8]

Kohn also points to other experiments that consider how we are taught to compete and how our perceptions are shaped once we become competitors. Experimenters have found that whereas cooperative individuals realistically perceive that some people are cooperative like themselves while others are competitive, competitive individuals believe that everyone is like them, that it's human nature to compete.

Even natural selection, Kohn argues, is a cooperative process. Contrary to the prevailing ethos of Darwin's survival of the fittest, nature is a proponent of collaboration rather than competition. Natural selection does not require competition; on the contrary, it discourages it, and survival generally demands that individuals of the same species, as well as those from different species, work with rather than against each other. Close examination demonstrates that animals cooperate with one another except in extreme conditions.

Kohn refers to the work of Russian dissident Peter Kropotkin, who was apparently the first to show that the animal kingdom is cooperative. In his 1902 book *Mutual Aid*, Kropotkin demonstrated that those periods in which there is heightened competition among species members are always harmful to the species. The tendency of nature to cooperate, Kropotkin said, although not always realized, is the constant message that comes to us from every landscape—mutual aid is the watchword. Kohn cites a number of more recent scientists who confirm not only that animals avoid competition but that their behavior is overwhelmingly characterized by the opposite—cooperation. He asks himself why the idea of a cooperative nature seems surprising to so many of us and concludes that it is harder to recognize than competition:

> Cooperation is "not always plain to the eye, whereas competition
> . . . can be readily observed," as Allee put it. Lapwings protect
> other birds from predators; baboons and gazelles work together to
> sense danger (the former watching, the latter listening and

smelling); chimpanzees hunt cooperatively and share the spoils; pelicans fish cooperatively. Indeed, the production of oxygen by plants and carbon dioxide by animals could be said to represent a prototype for the cooperative interaction that becomes more pronounced and deliberate in the higher species. None of this, however, makes good television. It is easy to ignore an arrangement that does not call attention to itself.[9]

Most of *No Contest* is devoted to a systematic unraveling of what Kohn calls the "four central myths" of competition: (1) competition is unavoidable; (2) competition motivates us to do our best—we would cease being productive if we did not compete; (3) contests provide the best, if not the only, way to have a good time; and (4) competition builds character and is good for self-confidence.[10]

When confronted with the question of eliminating competition in favor of cooperation, people tend to say that we can't possibly keep our world vital, exciting, and productive without competition, which is everywhere in our society—in business, politics, sports, and academics. Competition has helped us to scale great heights, we're told.

The effectiveness of competition as a mode for reaching specific, highly targeted goals—such as maximum profitability—cannot be argued. My experience has taught me, however, that we do better across a multiplicity of goals, producing greater, more lasting satisfaction, when we work in cooperative modes, balancing the needs of multiple stakeholders.

Another examination of the impact of a single-minded pursuit of competition arises from the work of Dr. W. Edward Deming, widely acknowledged to be responsible for the Japanese post–World War II industrial revolution, an American who only later became known as the "guru of quality" for American industry as well. In the 1950s, as the Japanese economy was rebuilt following the devastating effects of the war, Japanese goods were the butt of worldwide jokes. "Made in Japan" was a metaphor for cheap and shoddy. Today, decades of success by Sony, Toyota, Honda, and the rest have obliterated this memory.

At the request of the Japanese government, Dr. Deming worked to shift the world's perception of the quality of Japanese goods. He succeeded at this, in a very short time, largely by teaching the Japanese, through his total quality management system, to organize their workers in teams, to learn from the people on the factory floor, to drive out fear and banish exhortation, to do away with competitive bidding and arbitrary numerical targets, and to promote pride in craftsmanship. Japanese products were soon emulated throughout the world, and American companies were soon adopting the teachings of Deming.

For Deming, the practice of having employees compete with one another is "unfair and destructive." He believed that competition takes the joy out of learning and that annual ratings of performance and incentive pay "cannot live with teamwork." Not long before his death in 1993, he read Alfie Kohn's book and said, "We have been in prison from wrong teaching. By perceiving that cooperation is the answer, not competition, Alfie Kohn opens a new world of living. I am deeply indebted to him."[11]

Kohn is asking us not to do away with incentives or tests but to stop using them to determine a "winner." He is asking us not to do away with games but to remove the emphasis on winning. Our favorite team does not have to win every game, or win at all, to stay close to our hearts. Red Sox Nation endured its team's failures for eighty-six years.

I hope the spirit of Kohn's theory—that people in an equitable and cooperative setting will attain a goal with more efficiency and creativity than people in a competitive setting—is being cultivated at South Mountain. It's a spirit that cannot be imposed; it can only be offered. I hope it takes hold more and more. I hope we can accept our urge to compete but remain vital with a minimum of rivalry and contention, and I hope we will always dignify and honor our collaborative successes. They are key to the road ahead, and perhaps the collaborative model is key to changing the direction of our business future.

In *What Matters Most* author Jeffrey Hollender quotes Bob Massie, the former executive director of CERES (Coalition for Environmentally Responsible Economies), a group of forward-looking companies that have committed to consistent environmental improvement. Massie, says Hollender, thinks the most important cultural shift occurring now is a spiritual one that moves us from competition to cooperation. Massie says, "Take a walk through any airport . . . and look at the business books. They're all about teamwork and cooperation and making the firm more like a family. Look at the way the stakeholder concept is taking hold, all over the world, which is truly a triumph of cooperation over competition."[12]

Let us hope that this triumph continues to make progress. In the Emilia-Romagna region of northern Italy, where thousands of small businesses are the heart of an energetic manufacturing economy, businesses are finding that they can cooperate for a general, rather than an individual, benefit. Small businesses in the same line of work, which would normally be competitors, have begun to pool resources for greater buying power, for control over particular industrial processes, and for shared distribution. Everyone benefits, and small companies gain some of the advantages that are usually available only to large companies. These collaborative methods have helped the district become one of the world's most dynamic and prosperous economic regions. In his book *Making Democracy Work*, Robert Putnam says of the region, "A rich network of private economic associations and political organizations . . . ha[s] constructed an environment in which markets prosper by promising cooperative behavior and by providing small firms with the infrastructural needs that they could not afford alone."[13]

This might be the kind of economy Alfie Kohn would design if he were an economist.

Handshake Business

Collaboration and cooperation can also relieve us of the colossal waste of time that assigning fault represents. They can, even in today's litigious world, substitute handshakes and mutual understanding for the elaborate contractual protections that are necessary in a business environment defined by competition.

A client of ours once came to see his house as construction was ending. A large and important window just didn't feel quite right. It was exactly as shown on the drawings, but in real life the size of the panes didn't feel right in relation to the other windows. It was subtle, but it was so. I said, "Maybe you should live with it for a while and see if you get used to it."

His unforgettable reply: "I can get used to it. I can get used to almost anything. I can get used to a hatchet in my forehead, but why would I want to?"

Right. We don't want people to get used to what's wrong with our buildings. We want to make them right. We replaced the window. There is nothing gained by assigning fault. It's far better to understand ourselves as collaborators with the owner; we are each engaged in the project of making the best building we can. It's like author Tom McMakin says in *Bread and Butter*, his book about Great Harvest:

> We look at each loaf, and know that loaf is going to end up in someone's house, all alone[,] and that it won't matter to those people how good all the other bread was that day. The bakery will be judged on that loaf alone, even if 999 other perfect loaves were produced. This is true no matter what business you are in. A reputation is built day by day and loaf by loaf.[14]

As difficult as it may be for some to believe, an attitude of shared purpose can allow the handshake to be the principal mechanism for doing business. It wasn't so long ago that most commerce was transacted as a set of understandings shared among company, customers, and community, and I believe that a spirit of cooperation and commitment to multiple bottom lines can restore much of this lost understanding.

South Mountain has a three-page contract that, with only minor adjustments, has served us well for twenty-five years for projects up to $4,000,000. When new clients question the simplicity of this document, we discuss its purpose: simply to remind us, if we ever forget, about what we agreed to and what we're shaking hands about. They and we share the same small community, similar aspirations for the work we are undertaking together, and a sense of trust in each other. That's why we have agreed to work together. We're not going to wind up in a court of law; if we do, we all lose. And in thirty years, we have never been in court.

Rather than concern ourselves with what may be omitted from our contract and whether our language covers every eventuality, we just ask ourselves a few questions: What, exactly, do we wish to remember? Have we said that here? Is it a good reflection of our relationship and our mutual understanding? If so, it's a good contract. If not, no contract should occur. The purpose of the contract is not to guard against fleecing or fraud but to protect against failure of memory. We choose those with whom we do business according to our ability to achieve this degree of mutual trust.

People who do business in other locations have said to me, "That's all well and good if you can afford to choose your clients or customers. Working in a prosperous place like the Vineyard makes that possible." But this belies the fact that on the Vineyard, as in any other bustling economy, there is no lack of competition. The island contains a wealth of design and building talent; it's a miniature Silicon Valley of building expertise. The island's history of boatbuilding has attracted scores of craftsmen, and many have shifted to house building. When development and construction are booming and opportunities are great, professionals and tradespeople from Cape Cod, Boston, and New York swoop in. Someone's always offering a bargain, just like in Sioux City or Tallahassee or anyplace else. Wherever you work, if you believe in what you do and are committed to principles of quality and cooperation, you can't afford *not* to choose the work you are willing to do and the clients you are comfortable serving. We have found that when we elect to work with people we don't trust (or those who don't trust us), we are likely to lose both money and sleep.

Years ago we designed and built a house for a couple on the Vineyard. He is a writer and she is a judge. I remember the moment, sitting across the table from them, when it came time to address the construction contract for their house. We discussed the meaning of its contents. When we were through talking, I slid the contract across the table to them.

She took it first, flipped immediately to the signature page, signed, pushed it to her husband, and said, "Sign it, dear."

He turned to her, incredulous, and said, "Aren't you going to read it?"

She said, "Of course I will. Later. And if there's anything that needs to be changed, we'll call John and change it. Now sign it."

And he did.[15]

Coffee Break

I would like to suggest that our commitment to multiple bottom lines is expressed not only through wrestling directly with issues of accounta-

Coffee break in the old days on the rock under the swamp maple at the Chilmark shop. (Photo by author.)

bility and corporate culture, but also in a host of informal ways that define the overall feeling of the workplace. When we talk about serious business, we can't overlook coffee break.

Our office coffee break is a time for the designers, the business staff, and those who work in the shop (and anyone else who happens to be around on any given day) to sit and talk. Each of the job sites also has a coffee break that could be portrayed in the same way. But these days the office coffee break is the one I'm mostly at, so it's the one I know best. Our coffee break belongs to Jim Vercruysse; it's his creation, in a way. Jim is one of our owners and has been running our woodworking shop for nearly fifteen years. We always had coffee break (I think) and we didn't always have Jim (I know), but some people seem to become symbolic of things they didn't actually invent. That's the way it is with Jim and coffee break. He makes sure it happens (even though it happens when he's not around). He loves food and he loves to cook, so he is the inspiration (although not necessarily the provider) for the good food that sometimes shows up at break.

Lunch at South Mountain is similar to coffee break, another time of gathering. In fact, one of our newest employees, Betsy Smith, said to me at her first evaluation meeting, "I was a little worried to hear that most people eat lunch here. I need my *space*. When I worked at Honeywell— not that I didn't like my job—but I just had to get out of there. And I never went to coffee breaks at Honeywell; that's where the old guys would sit around and complain about how good it used to be. But here you hate to miss it. It's so much fun and it's where stuff happens. I love it."

Betsy's work is not glamorous—she types, she answers the phone, she files, she organizes, she does the mail. But it's tremendously varied and important. She buys flowers for clients and presents for new babies, she assembles slide shows, she helps people find what they're looking for, she helps people find other people, she makes all kinds of arrangements, she handles our insurance accounts—she's a critical hub. Invariably she has multiple active projects going. She has told me that this is her dream job. Her job description, and everyone's, for that matter, should also include "implementer and evaluator of multiple bottom lines." Betsy said that before we hired her, one of her goals had been to find a job where she was actually proud of the product. Now she contributes to the making of beautiful houses and the doing of good things in the community—that's what she *really* does, while she's doing all those other particular tasks.

Anita Roddick of the Body Shop once said that the way she evaluates the character of a business is by looking at the lavatories and the eating place. If they are dull and drab, it means something. If her criterion for business evaluation has merit, we would get a decent mark. The South Mountain shop and office has the same craft in design and construction as our homes, serving well as the center of a community and, more to the point, as the home of our coffee break.

At coffee break and at lunch the conversation can go in any direction. Political talk. Sports talk. Talk about movies, people, trips, art, music, and Vineyard events. Talk about South Mountain business, South Mountain people, and hot local issues. Some of the most important communication in the company happens at these times. Mostly it's a time to converse in an unstructured setting, for breaking and emerging news to be communicated, and for trying out new ideas before they're fully cooked.

It's a time for figuring out what's next.

One day I was preparing to leave for San Francisco. At coffee break,

Greg Small said, "When are we gonna get the corporate jet?" Comments followed about Ken Lay and Enron, the Sunday-afternoon crowd that jets into Nantucket to play a round of golf, "cheap" time-sharing jets, and so on.

I said, "Corporate jet? We haven't even gotten it together to have a company car on the mainland."

"Right," said Jim. "We need a corporate Jetta."

Coffee break is an attitude. I think of it as our expression of pleasure with the workplace we have created and who we are. Coffee break as an institution may seem trivial when set next to the complex and abstract concept of multiple bottom lines. But it nurtures something nontrivial in our process and business culture. It is an informal declaration of the mutuality and shared decision making at our core. It is a place where multiple bottom lines can mingle, relax, breathe, and find their own happy medium.

Multiple bottom lines lead to the notion that work should be a place where people can "work to their heart's content," as Sony founder Masaru Ibuka wrote in his original purpose statement for his new company. Work is a place where people should *be* content. It should be a place where people are proud of what they make, like Betsy, and proud of how they conduct themselves. That's what we're aiming for.

Nearly thirty years after our seat-of-the-pants beginnings, we are still small enough to stay closely connected to our roots, to do business on a handshake, to all gather in one small room, to know each other as people and not only as coworkers, to recognize one another as collaborators in pursuit of multiple goals. Living the language of our mission, goals, and purposes, and learning to collaborate together, has shaped a dedicated, skillful, compassionate body of decision makers. Nobody's getting rich, but we are living comfortably doing the work we enjoy in the location of our choice. *All* of us are able to make good livelihoods *because* no one of us is getting rich. Perhaps we have unconsciously internalized the wisdom of Chinese philosopher Lao-tzu, who said, in the third century BCE, "He who knows he has enough is rich." We are rich in multiple bottom lines.

Pete reading a poem at a company celebration—he usually writes them on a shingle. (Photo by Derrill Bazzy.)

Fellow Workers

In early December of 2002, one of our employees was arrested, for the second time in less than a year, for driving while intoxicated. He had been with us many years before, had left to go to sea for a while (he is a skilled and experienced mariner), and had returned. He's a meticulous carpenter and has always been well liked by his fellow workers. He had only recently gotten his license back from the first incident. Now it was clear that he was really in trouble.

Several of his friends on his crew, along with another close friend in the company and one of our owners who had had alcohol problems himself decades ago, decided to meet with him. Peggy MacKenzie, the chair of our personnel committee, pulled the meeting together. It was held in our conference room at the office.

He told me later that he walked into that room in a state of dread. He felt doomed, knowing they were going to tell him what a screwup he was and that they didn't much want him around unless he could get it together, which he obviously couldn't. But something different happened in that room. He said he walked out of the room, about an hour later, feeling loved. That probably wasn't easy for him to say to me, but it flowed out as naturally as could be. He walked in with dread and he walked out loved. He checked into a detox facility within a few days, with assistance from those same friends. A year later, to the day, the same five people took him out to dinner to celebrate his first year of sobriety.

I was lucky enough to spend a day skiing with him the following winter, and he told me about the dinner. As he talked, I realized that he had become a different person. I had never known him this way before. Gratitude and emotion filled his articulate account of the changes in his life. I thought to myself, "Look what happened here—this is astonishing. *Now that's the way to run a company.*" But the thing is, we're not running it that way. Our company didn't make that change happen; his fellow crew members, combined with his own strength of character, did. It had sprung from the personal values they all bring with them to work, not from any kind of management decision. Maybe a work environment that is more than a job encourages that kind of response. Maybe a company culture that encourages sympathy and cooperation helps inspire the kind of camaraderie that leads to such kindness. As Jim Collins says in his seminal study of enduring visionary companies, *Built to Last*, "You do not 'create' or 'set' core ideology. You *discover* core ideology. It is not derived by looking to the external environment; you get at it by *looking inside.* It has to be authentic. You can't fake an ideology."[16]

You can't fake it. It's there or it's not. It's supported or it's not. You discover what it is. Those cornerstones in the pile—they really have to be there, you really have to find them, and you really have to place them in the wall.

Path to the beach. (Photo by author.)

$\lceil 5 \rceil$

COMMITTING TO THE BUSINESS OF PLACE

With ten seats maximizing the passenger seating in the small cabin, this plane is built for neither comfort nor speed. The twin engines of the Cape Air Cessna 402 putt like an old Evinrude outboard. Roughly fit aluminum is fastened with rows of rivets and screws and the wear of years ripples across the wings—no shining true surfaces here, just sturdy no-frills construction and practicality.

I'm alone in the plane with the pilot; nobody else is heading from Boston to Martha's Vineyard on this sparkling clear April afternoon. This is the off-season. The "shoulder" season has not yet begun. We motor out toward the runway. The little Cessna scampers among the 747s and DC-9s like a Border Collie herding buffalo. The pilot briefs me and we take our place in line. USAir takes off. Delta follows. Continental. USAir again. United. Our turn now. We leave the ground.

From above, Logan looks gritty with the debris and disorder of renovation and construction. The city itself looks like a fully assembled puzzle, with pieces fit together miraculously. We turn south. I'm mesmerized for a time by the great wind turbine on the tip of the Hull peninsula, gracefully powering small-town life, and by the juxtaposition of the simple turning of those blades with the complex operations of the airport, city, and Big Dig just behind us. We make our way to three thousand feet. The plane bounces and buffets on changing air currents like a mountain bike on dirt road corduroy. The brilliant blue of oceans, inlets, lakes, and rivers mixes sweetly with the pastel greens, reds, and browns of fields, cranberry bogs, and woods.

As we approach Buzzards Bay, I see the Vineyard ahead and the Elizabeth

Islands strung out to the west, beating a path to little Cuttyhunk, the out-ermost. Before we reach the Vineyard Sound, the pilot begins to drop alti-tude. We pass over Woods Hole at twenty-two hundred feet. The Vineyard shoreline ahead is crisp and complex, each crenellation recognizable. The North Shore bluffs are sullied by an occasional house. Then more houses come into view, scattered like chess pieces swept off a board. But mostly I see woods and fields and water. Those who fly over the Vineyard for the first time are often amazed at how undeveloped it seems, how much space is left untouched.

We're close to home and flying low. As we descend, a more detailed picture comes into view. The small towns at the island's edges are clus-tered around steeples on shore and masts in the harbors. The wooded middle is broken up by the gravel pit, the high school, the hockey rink, and the small industrial park. The airport is just ahead. We hit the ground, bounce back up, and settle down for good. The plane slides to a halt. I unfasten my belt, crouch and shuffle to the rear, duck to exit, and take the few steep steps down. Home again. Standing on this ground brings a special comfort.

It wasn't always that way. When my partner Mitchell, my wife Chris, my four-year-old son Pinto, and I moved to the Vineyard in the 1970s—from the hills of Vermont and the mountains of British Columbia before that—it felt like an alien, one-dimensional landscape and a parochial cul-ture. But, happily, our work has kept us here long enough for us to begin to know this place.

What Kind of Community?

As we came to each cornerstone and as our company became more con-nected to our community, I began to think differently about the products of our work. My affinity for old structures led me to wonder, as we made new ones, what their fate would be. Would the buildings endure and be used for centuries like the venerable barns I loved? Would the summer houses we made, like the hallowed old shingle-style houses of the Cape and Islands, become family compounds that would be enjoyed for gener-ations of extended families? The desire of my twenties—to live in another time—evolved into a different quest: to craft buildings that were sturdy, timeless, and beloved, and to imagine how people would live in them and care for them. Emphasis shifted from past to future. Our houses began to have an eclectic character that linked each to the others. They did not mimic the old; in fact, they were intended to be buildings for tomorrow. It has been said that the first rule of intelligent tinkering is to save all the parts. Salvage materials and solar panels—we saved the old parts and mixed in the new.

Like a hound picking up a scent, I seemed to be homing in on a new way of thinking about the future of our buildings, which is less about the structures themselves and more about the places they will inhabit. I was becoming attached to the Vineyard, and as I began to sense that my future was tied to the island's, I became increasingly invested in its fate. We know enough to make buildings that will be around for our grandchildren's grandchildren, but we don't know what kind of world theirs will be. What kind of landscape will surround them? How will people make a living and how will they govern? How will their energy be supplied and how will they move around? Will they feel safe and secure, satisfied and

The Allen Farm in the early days of South Mountain Company—the shop is on the right. (Photo by Peter Simon.)

fulfilled, and will they treat each other with kindness and civility? Our business inquiries were leading to a new understanding: that an essential aspect of commerce is building community. We had assumed, in the '60s, that the choice was between accepting an existing community (or society) and creating a new one; now it was dawning on us that we could *partici-pate* in designing the future of the community we'd joined, along with the buildings that would be a part of it.

What kind of community will it be? This much we can know: it won't be the vanished American past I had been so taken with. Many of us wish for those strong, close-knit communities and simple ways of life, but we also demand qualities that were often missing in the bucolic past—diversity, tolerance, equality, comfort, innovation, and excitement. We won't go back, but what would forward be? I think back to our social endeavors of the '60s and '70s, when we had distanced ourselves from the mainstream culture. Today, in 2005, I think we're still working on the same project, but with the refinements of age and new understandings, a sense of cultural integration rather than isolation, and, most important, a different relationship to time. Urgency has given way to determination, and our forays among the remains of the past have evolved into purposeful commercial endeavors with an eye to the future.

The Texture of Community

In our early years we had a small fleet of character-rich and mechanically challenged vehicles, including a '55 Pontiac Stratostreak, a '65 International Harvester flatbed, and a Dodge panel truck that had previously been abused by the local utility. All needed constant attention. It was not rare to find myself in Ken Dietz's radiator shop. When you said good-bye to Ken he always replied, "Fly low." Stay under the radar. Ken died years ago and there is no longer a radiator shop on the island, but Martha's Vineyard remains dependent on many small businesses like his.

Businesses of every sort are woven into the fabric of this community. Bait and tackle shops mix with upscale galleries and antiques stores. Tailors, truckers, plumbers, and pet groomers all have their places. There are a slew of realtors, a boatload of builders, more bakeries than banks, too many T-shirt shops, and one cobbler. Farmers and fishermen still harvest from land and sea. Boatbuilders still hew to a line. These small family businesses are the ballast of the community. They create a kind of gravity, holding us together, stabilizing us.

People often ask why anyone would devote so much work and effort to a place that is widely perceived to be just a playground for the wealthy. While it surely is that for some, it is many other things as well. For me, four distinct elements define the texture of the Vineyard community: it is the place I know; it has "fifth migration" qualities; it is socially complex; and it is in "fair" condition.

The Place I Know

I've lived in fourteen different towns and cities in my life, but I have lived on Martha's Vineyard for more than half of my fifty-five years. Although I will always be a "washashore," my children grew up here (one was born here), my grandchildren are now growing up here, and I am beginning to comprehend this place, just a little, through long association.

As a kid I was fascinated by geography. I pored over relief maps and atlases, wondering what mysterious places lay hidden, memorizing place-names and populations. In my late teens and early twenties I loved nothing more than being on the road. I still wonder what's around the next bend and I still love geography. So how did a wanderer like me get hooked on a tiny island? The Vineyard seems small and limited, but after thirty years I still find new dirt roads I've never been down, new trails to walk for the first time, and new vistas to take my breath away. I've come to think that maybe to know many places we need to deeply know at least one.

Our relationship with the Vineyard is rich and complex. If Chris and I were to move soon, we might be able to have as thorough a relationship with one more place in our lifetime, if we were lucky. But that is not likely. Here, we've watched our children and their friends grow up, and we've watched the children have children. We see them in the streets, at the grocery store, and at the movies. This experience is neither replaceable nor replicable. Most of the people of South Mountain are as deeply connected to the Vineyard as I am. Some were born here. Others have thrown in their lot here. It's the place that we know.

Fifth-Migration Qualities

Author Peter Wolf says that Americans are in the fifth national migration of our country's history.[1] From 1600 to 1785, discontented Europeans and enslaved Africans settled here. From 1750 to 1890, the poorest Americans pushed westward. From 1820 to 1920, millions of job seekers moved into industrial cities during a century of urban concentration. Between 1930 and 1990, one hundred million people drove out of town and created the suburbs. The fifth migration, which began in the '70s, is a dispersal—across the country, Wolf says, millions are streaming into communities distinguished by physical beauty, abundant recreation opportunities, clean air and water, and relatively few social problems. I would add to Wolf's list the following attractive qualities: tolerance, diversity, and educational opportunity.

The towns and cities that are the object of this fifth migration are the most desirable places to live. As the U.S. economy shifts from manufacturing to information, many cities, towns, and regions are welcoming the rapidly expanding "creative class." Sociologists Richard Florida, in *The Rise of the Creative Class*, and Paul Ray and Sherry Ray Anderson, in *The Cultural Creatives*, hypothesize that this identifiable group now includes forty to fifty million Americans. Florida defines the two parts of this new class. He says that the core of it includes scientists and engineers, university professors, poets and novelists, artists, craftspeople, entertainers, actors, designers, and architects, as well as the thought leadership of modern society, including nonfiction writers, editors, cultural figures, think-tank researchers, analysts, and other opinion makers. The other part includes "creative professionals" who work in a wide range of knowledge-intensive industries such as high-tech sectors, financial services, the legal and health-care professions, and business management. More and more businesses are locating to the places where these people (the resource businesses need most in the information economy) want to be. Centers of creativity are developing in such locations throughout urban, suburban, and rural America. Florida calls them "creative centers" and says that they are thriving, but not for such traditional economic reasons as access to natural resources or transportation routes or tax breaks and business incentives. Rather, they are succeeding largely because people want to live there. He goes on to say, "Perhaps the greatest of all New Economy myths is that 'geography is dead.'"[2]

Along with being a resort community, Martha's Vineyard has all the qualities listed above. It is safe, beautiful, and culturally rich. It maintains strong characteristics of community. Many live here for that simple reason; some seasonal residents return year after year as much for the community as for the beauty. My friend Jamie vacations here each June with his family and says the first thing he does when he arrives is to get a local paper and catch up, because our community is more real to him than his own! He says, "You face all the same issues we do, but there is a tangible sense that you can talk about them as a community and work together to do something about them at a scale that makes accomplishing something feel possible."

Whether we can solve the issues we face as a community is always in question, but at least they get fully aired out. The Vineyard has those *fifth-migration qualities* that make it an attractive place to be.

Social Complexity
It's interesting to discover that the Vineyard is also *socially complex*. A friend who grew up in a poor rural area says he's from a place "where most of the pickup trucks were up on cement blocks and the houses all had wheels." Before my family and I moved to Martha's Vineyard, we lived in several rural areas that fit that description—down-at-the-heels regions that had seen better days and saw little opportunity to restore what had once been vigorous rural economies driven by farming, fishing, ranching, or logging. The Vineyard is cleaner, almost whitewashed; it has lots of open space mixed with spruced-up towns, due to the thriving tourist economy that has replaced the declining rural farming and fishing heritage. But this new economy, ironically, is helping an agricultural and craft-based revival occur as the demand for fresh food and authentic experience grows.

Although it looks pretty spiffy, it's not so rosy. The junk-cars-and-plastic-toys-in-the-yard look may be less prevalent, but it's just a matter of space. In most poor rural areas, there is room to spread out, so the stuff of peoples' lives is more visible. In resort communities with high land values, space is at a premium. The Vineyard can more aptly be compared to a city, which has poor neighborhoods dense with people and other, more sparsely populated areas where the well-to-do folks live and/or vacation. This is not a wealthy county. Of the other eighteen counties in Massachusetts, half have higher median incomes than the Vineyard, and none (except the farther-out island of Nantucket) is a more expensive place to live. The year-round population is not so affluent.

The social mix compares to that of cities as well. The high profile of some of the wealthy and celebrated can distort the picture. Along with them, there is a large service class that caters to their needs, there is considerable poverty, and the high real estate values make affordable housing scarce. The Vineyard has a middle class made up of small-business owners, tradespeople, professionals, and white-collar workers, although the housing crisis threatens even this group. There are ethnic enclaves (Native Americans, African Americans, and Brazilians),[3] a growing population of retirees, and a thriving arts scene. The Vineyard is also seeing an influx of young families that depend on the information economy for livelihood and therefore are not tethered to any particular place. They are attracted by the fifth-migration qualities that the Vineyard shares with small cities like Portland, Oregon, and Portland, Maine, and a host of college towns, state capitals, and resort towns across the country.

The profile of seasonal residents is changing, too. We have built a number of houses for people who originally imagined them as summer houses, only to have their Vineyard house become their primary home because they find the way of life the island offers appealing. Many have sold the now oversize suburban house where they raised their families, rented or bought an apartment in the city, and now spend a larger part of each year living on the Vineyard. Some have begun to vote here, because their vote may have greater impact in local elections, which have such a direct impact on their adopted community and therefore on their lives.

What's happening to Martha's Vineyard may presage what's in store for significant parts of America as our economy and demographics shift. More places will develop their fifth-migration qualities as more people have enough financial resources and job freedom to relocate. But desirable places all share at least two vexing problems: (1) outmigration forced by dramatic real estate appreciation and housing costs, and (2) changes to the character of the community resulting from the influx of people who seek to enjoy qualities that their arrival increasingly imperils. We all know places that have suffered this fate.

"Lace" walls along ridges like this one were built with space between the stones so the wind wouldn't blow them over. (Photo by author.)

I think of islands as laboratories, and the Vineyard is a good one for rec-ognizing, testing, and working to enhance the connections among small business, the built environment, and community, and for realizing the potential of commerce, local government, and nonprofits to collaborate for the common good. The physical isolation may allow us to see things more clearly, because boundaries and limits are so well defined. We are dealing with spaces we can understand. Islands are semiclosed systems. When you get off the boat or the plane and set foot on the Vineyard, you know that you are in a place that is defined, that has limitations. The social complexity combines in interesting ways with the fixed bound-aries, creating conditions for innovative problem solving and community initiative.

Fair Condition

From an environmental perspective, the Vineyard is like the old '55 Pontiac Stratostreak I drove when we first arrived: in *fair condition* and decent running order. It has been capably preserved, always garaged, but not yet restored. It needs attention if it's to keep running.

Islanders have always been aware that our unique environmental qualities are strong suits, tied directly to the character of the community and
our economic vitality. Some of our early seasonal residents were
visionary conservationists. In the 1890s, a Harvard geologist created
Seven Gates Farm, putting seventeen hundred acres into a permanent
trust that provided for just thirty homesites and no private landowner-
ship (those who have built houses there lease the land). It was carefully
planned; each house had to touch a steel stake that he drove into the
ground on its site. In those days, Seven Gates was a long day's carriage
ride from the ferry in Vineyard Haven. These properties (although you
can't even *own* one) have become some of the most expensive on the
island because they are in an uncommonly well-protected location.
There are still a few unbuilt properties in the trust, and there, more than
one hundred years later, you can still find the steel stakes; in fact you
must, because your new house must touch that stake.

The island has a robust tradition of land conservation and stewardship.
Private nonprofits have been buying and managing local land for
decades. The Nature Conservancy, a major landowner, conducts long-
term projects to restore ancient sand-plains environments. In 1986, island
voters created the Martha's Vineyard Land Bank, an unusual public body
(there are still only a half dozen in the nation) that collects a 2 percent
transfer fee on land transactions. The take from this has averaged roughly
$8,000,000 annually the past five years. The funds are used to purchase,
hold, and manage property for public access and environmentally appropriate use. The Land Bank now owns roughly two thousand acres,
including many beach accesses, and is in the process of creating a neck-
lace of properties that, over time, will make it possible for one to reach
most of the island by foot, horse, or mountain bike without using roads.
The organization is neither a sanctuary program nor a park system;
rather, it is a middle ground in which conservation values are balanced
with public enjoyment, and sophisticated management secures the land's
permanent stability.

At the same time, "people conservation" has taken its place alongside
land conservation as the community has united to solve the affordable
housing crisis. Without a stable supply of affordable housing, people are
forced away and the community is disrupted. We are beginning to under-

stand that "continuity of generations" (as local ecologist Tom Chase calls it) is essential for keeping a place whole—it's what keeps stories and traditions alive, maintains a population with a good understanding of the land and the climate, and provides a window to the past and the door to the future.

Sprawl has not yet gotten out of hand. The villages still feel like villages. Recently I was speaking to a group of public officials and concerned citizens who wanted to begin to solve the affordable housing problem on Mount Desert Island in Maine. A local builder, Eric Henry, described their situation with a comparison:

> Our island is like a salt marsh. The tide comes in, brings water and nutrients and plankton and pollution and garbage. The tide goes out, leaves what it does, the marsh absorbs what it can, and then it rejects the rest. That's like our summer. Visitors bring money, culture, traffic, pollution, you name it. We take what we want and reject the rest. But we have to remember to do the rejecting.

Yes, we must make good choices, and in many cases we have. In others we haven't. For example, large-lot zoning initiatives adopted in the 1970s led to the unintended consequence of partial suburbanization of a rural landscape. Inadequate transportation planning and funding has led to serious traffic problems. But the Vineyard remains in fair condition. Restoration is conceivable.

Basil

Sometimes random incidents, apparently minor, become emblematic, helping us to see our communities and ourselves in fresh light.

One day while my partner Mitchell and I were in the process of building our first Vineyard house, we, with a helper, were raising a timber wall. The wall was too heavy for the three of us. We raised it to shoulder height but could go no farther; we could rest it there, but we hadn't the strength either to go up or to let it down easy. Stuck there, we considered counting to three, letting it go, and jumping clear. At that moment, the telephone repairman arrived to fix our job-site phone. He saw what was up, slammed on the brakes, hopped out, and lent a shoulder. It was just enough—the wall rose past the pivot point and settled into place. Once it was secured, we breathed deeply, congratulated each other on our good fortune, and thanked our redeemer for his timely arrival and spirited participation.

We were new to the area then, and that was the first time I'd met Basil, the telephone repairman. I saw him over and over after that day. For years he never failed to say, "Hey, got any walls to raise?" For him, the opportunity to help us out of a jam was memorable; for us, he became an example of small-town camaraderie and a reminder that our business can be, to our community, what Basil was, at that moment, to us. Pitching in to help each other, and our town, and our region, makes us more alive and present. It make us part of the stories of others.

Local Commitment

As South Mountain has developed and our bonds with the Vineyard have strengthened, we have made several commitments to ourselves regarding the work we will do and the work we will not do. One of these—the commitment to work only on Martha's Vineyard—is central to this chapter. Its emergence led directly to our fourth cornerstone principle: committing to this island in order to promote local economic stability and to strengthen community. Here's how it happened.

In 1995 we expanded the company's reach. South Mountain became a partner in a new practice called ARC Design Group. I had met the other two partners, architect Bruce Coldham and systems engineer Marc Rosenbaum, years before. Our first work together was organizing conferences about green building as board members of the Northeast Sustainable Energy Association (NESEA) in western Massachusetts. NESEA brings together renewable energy and green building advocates and practitioners from all over the Northeast. It has been a remarkable spawning ground for professional friendships, business associations, and collaborations—a true community of practice.

In 1993 South Mountain was asked to design the Wampanoag Tribal Center on the Vineyard, which was to be the first building the Wam-

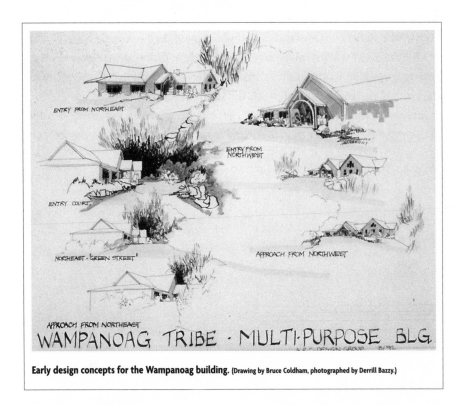

Early design concepts for the Wampanoag building. (Drawing by Bruce Coldham, photographed by Derrill Bazzy.)

panoags had built, as a tribe, in three hundred years. The project was beyond our capabilities, so I asked Marc and Bruce to collaborate with us. They did, and the partnership emerged, with the specific purpose of bringing integrated planning, architecture, engineering, and construction management to New England institutional, commercial, and multifamily residential projects in which the client had a particularly strong commitment to environmental sustainability. While South Mountain's residential practice on Martha's Vineyard remained our bread and butter, ARC Design completed several successful projects during the second half of the '90s.

In 1999 we met to discuss expanding the practice and specializing in cohousing, a particular kind of residential neighborhood development. On the long drive home from that meeting I wrestled with the idea. I was intrigued but also troubled. It was an opportunity to do work that fit our company values, but when I examined the benefits and detriments, the benefits kept coming up short, no matter how I spun it. The detriments—significant travel (time away from family) and disparity within the business (those who worked off-island and those who didn't)—seemed great. The more I thought about doing this kind of development, the more unappealing it became. Ultimately, I came to believe we couldn't do the job with the kind of depth and attention to detail to which we aspire.

Successful development work, in my view, relies on deep understanding of people and place. You have to *be* there. After a quarter cen-

The lobby of the Wampanoag Tribal Center. The plants in this room convert the building's wastewater to useful nutrients. (Photo by Derrill Bazzy.)

tury working on the Vineyard, our local knowledge is still increasing. Even here, there is so much more to learn. Sometimes we still slog through the tangles of information like a short-legged dog in deep snow.

Tom Kelley, CEO of the industrial design firm Ideo, says in *The Art of Innovation*:

> Inspiration often comes from being close to the action. That's part of why geography, even in the internet age, counts. New ideas come from seeing, smelling, hearing—being there. . . . It's . . . why people still go to museums, to be inspired in the presence of original artwork, though a digital image may be easily available on their home computer screen. Asking questions of people who were there, who should know, often isn't enough. . . . It doesn't matter how many astute questions you ask. If you're not in the jungle, you're not going to know the tiger.[4]

Every region has distinct landscapes, microclimates, and local cultural practices that influence design and building practice. As I collected my thoughts about practice, locale, and business, I began to come to a provocative idea about time: What if we, as a business, decided to confine our work to the Vineyard, for the next hundred years at least, to help make it the best place we can imagine, the kind of place we'd want our grandchildren's grandchildren to live in? Could we become a part of that "continuity of generations" that Tom Chase talks about? If we imagined our eggs in that one basket, over that time perspective, our work might take on a different cast. In fact, I thought, with some excitement, "What if we went further and made a pledge to maintain the buildings we make in perpetuity?" Stewards of our buildings, the land, and the community—it wasn't *so* far-fetched, was it?

I was anxious to share these new thoughts with my co-owners. When I did, there was agreement, and our deliberations led to this cornerstone commitment, as a company, to limit our work to the island of Martha's Vineyard, with the exception of educational work beyond our shores. That covenant defines and guides us today. We get many requests to work in other places. In the past, each inquiry was a seduction that had to be evaluated. Now we are clear: if we can be educationally helpful in another community we will try to accommodate, but when it comes to actual design and building projects, we keep to the Vineyard. We're staying in our jungle.

Beyond Global to Local

After reading an article in the Hartford, Connecticut, newspaper about the closing of a local plumbing supply company, my friend Jamie Wolf was moved to write the following letter to the editor:

> Like the contractor in your story about the closing of Carlton Supply, I learned a lot sitting across the counter from Mike as he patiently explained the nuances of the plumbing in the old Lincoln Street two-family house I had just bought and was renovating. It was 1978 and I was one of the many who, in their 20s and early 30s, found the opportunity in Hartford's aging neighborhoods to own, and often improve, homes with architectural character in richly diverse areas. When I created a business in remodeling I continued to benefit from relationships with people like Mike in businesses like Carlton Supply. Most of these businesses closed up in the economic slump of the early '90s or were driven out by the arrival of the big boxes. I'll never forget the day I called a local millwork company looking for an old house's big fat door trim. "What neighborhood's that in?" the fellow asked, and when I told him he replied, "We've got the cutter to mill that. We made the original." That was my lesson in the equity of local knowledge and irreplaceable community resources. I can't get that trim anymore; they're gone. The loss of Carlton Supply is one more such loss to the community. The next young urban homesteaders will have to figure it out for themselves. Alas for Hartford. The heart and wisdom offered by folks like Sid and Mike and Junior are the genius of the place. The city will more than miss them.[5]

Jamie's letter touches on many of the essential characteristics of a stable community economy: familiarity, local knowledge and production ("we made the original"), diverse small businesses, and the desire and responsibility to be helpful beyond selling a product or service. These valuable businesses, which are the heart of local economies, can be lost, just like species and forests and soil, if we don't protect them, honor them, and nurture them.

Every community has businesses and institutions that, if they were gone, would fundamentally change the character of that community. I can think of many here. I cannot imagine the Vineyard without Trip Barnes's colorful trucking company (or Trip himself, for that matter, who presides with humor and passion at many of our charity auctions), or Reynolds Rappaport and Kaplan, the legal firm that manages to do both public and private work and stands as a symbol of integrity. Life would be different without Gannon and Benjamin, the wonderful builder of wooden boats that routinely draws five hundred to a thousand people to celebrate the launchings of its extraordinary vessels. There's no doubt we would live in a lesser place if not for the presence of Chilmark Chocolates, the public-spirited candy maker that provides a homey working atmosphere for those with disabilities, supports many local causes, and fattens us up joyously each holiday. How about our fiercely independent bookstore, The Bunch of Grapes, which author William Styron calls "the best bookstore in America"? Founded by Ann Nelson in 1975, the bookstore has now been passed on to her son John, and it maintains the selection, the personality, and the vitality it has always had.

There are many more I could mention; these few are just particularly representative.

Recently I visited Brattleboro, Vermont, for two different gatherings. At each I asked, "What businesses define this town? Which are the ones that if they closed their doors tomorrow would immediately diminish the quality of life here?" Each gathering comprised only a small sampling of people (twelve people at the first, eight at the second), but there was a surprising consensus. Two businesses appeared near the top of everyone's list: the Brattleboro Food Co-op and Brown & Roberts Hardware. The Food Co-op, down at the end of Main Street, is a modern-day supermarket with a huge inventory of fresh high-quality food, a lively café, bulletin boards with notices hanging off its edges—in short, it's the village square. It's well lit and intuitively organized, and the

Barnes Trucking—it's one of a kind! (Photo by Randi Baird.)

merchandise is thoughtfully displayed. Much of the food is produced locally. The people who work there are friendly and engaged. The shoppers don't seem to be in a rush to get home.

Brown & Roberts, up the street, has a drab exterior with a plain sign in block letters. Inside, however, this family business, which has served Brattleboro since the 1800s, sings with life. It is remarkably well stocked, from an entire section of kerosene lamps and accessories to an up-to-date offering of laser levels. The people on the floor have real answers for all questions. I went to the counter and said to the woman there, "I need two things: a bulb for a Mini Maglite and some history."

She replied, "I can give you both." It turns out that Linda and her husband are the current owners, descended from the original founders. She traced the family history with the business as it has evolved and the several locations it has occupied. There was feeling in the way she spoke; her long-term connection with those who had come before her and with the town and the region where the company has operated for more than a century has meaning for her.

These two businesses are as different as a laptop computer and a Royal typewriter. One is clearly a product of the times and the other is a remainder from a different era. But in fact the two stores share many similarities—a strong sense of familiarity with their customers, employees who are there because they want to be, dedication to providing high-quality service and a wide variety of the best available merchandise, and a commitment to community. All vital places need such enterprise.

Maintaining core businesses is critical, but it's only a beginning. As the South Mountain local commitment has developed, I have become curious about the potential of local economies to serve as an antidote to the negative consequences of globalization. I don't propose that we try to bottle up the escaped genie—globalization is here to stay and brings benefits along with its problems. It connects people worldwide, helps us understand each other's situations and differences, and provides a framework for solving global problems. But the injustices are legion, and they may be difficult to correct. As globalization unfolds, local communities and economies are endangered by mobile capital and corporations that lack deep connection to the communities in which they operate. The steady corporatization of our lives is like the fabled giant squid attacking Captain Nemo's *Nautilus* in *20,000 Leagues Under the Sea*: there's always another tentacle slithering through to make more trouble.

While we struggle to correct that which we cannot fully control, we also can invest in that over which control is more direct: the process of localization. If the people who *make* the decisions are the people who will also bear the *consequences* of those decisions, perhaps better decisions will result.

Subsidiarity and Federalism

The idea of supporting local economies is to increase local self-reliance, which requires a commitment to satisfy as many needs as possible at the local level and to reduce reliance on imports and exports. If we do as much as possible as close to home as we can, we will create a supply of good stable work, a reservoir of strong community engagement, and the economic multiplier effect that comes from money being spent over and over in the same place. A good frame of reference is the principle of subsidiarity, which proposes that matters should be resolved, decisions made, and responsibilities accepted at the lowest possible level of organization, the level closest to the people affected by the decision. The term is used mostly in reference to governments and churches, but it extends comfortably to business and social organizations as well. It is simply a preference for the most local decision-making and economic activity that is adequate and appropriate for a given task. If we apply this to Martha's Vineyard—or any other small town or region—it means that the more food, energy, and other essentials we produce, the more investment we access locally, the more control we exercise over our economic institutions, and the more political decisions we make at the local level, the better our community life will be. Our community will be more responsive to its inhabitants and the inhabitants more responsible for its fate.

Making choices with the subsidiarity concept in mind may allow us to benefit from the explosion of knowledge and products in the global economy, while at the same time we continue to work for vibrant and democratic community economies. Embracing subsidiarity is an argument not for minimal government or dogmatic small-scale but rather for a society of appropriate scales in which we, as individuals, are influential in as many of the decisions that affect our lives as possible. The same reasons that make it desirable for people to own and control the companies they work in—because people should have the right to determine their own destiny, because those who do the work should share the bounty, and because being stakeholders makes people more effective and responsible decision makers—seem to apply to economies as well.

Michael Shuman, in his book *Going Local*, points out another virtue of subsidiarity. He says that local failures are usually smaller, less catastrophic, and easier to fix than those that occur on a larger scale. There may be more opportunity for influence, too, on the part of those affected by decisions, because it's harder for decision makers to hide from their mistakes when they're part of a community. As Shuman points out, however, this doesn't mean that local decisions are "necessarily efficient, fair, democratic, sensitive, creative, or disaster-proof."[6] But there is more incentive for responsibility and transparency when things are decided and conducted close to home.

Charles Handy uses the word *federalism* to express similar ideas.[7] Federalism, or the distribution of power between a central authority and constituent units, is a familiar political concept. There is constant discussion and adjustment in government about what is appropriately federal jurisdiction, what is under the purview of the states, and what is properly local. The concept of home rule, or the right of a town or city to enact laws that are primarily municipal and do not violate state law, is well understood. However, Handy believes that democracy must become far more local if it is to work.

Handy also believes that federalism is an inevitable form of organization for business. Applying federalism to commerce recognizes that businesses are communities and that citizens must have a voice in their future. In my town and in many others, every time municipal officials meet, the meeting must be open to the public. Laws, regulations, and initiatives are enacted by residents sitting together at a town meeting several times a year. Everyone has a voice, whether they choose to exercise it or not. If I want to invest more, I have that option. There's a place for my voice.

Subsidiarity and federalism are not absolutes and they do not work in one direction only. Martha's Vineyard has six different towns on one small island of less than a hundred square miles and a population of sixteen thousand. That means we have six town halls, six police departments, six school committees, six planning boards, six boards of health, and so on. Does this make sense? Mostly it doesn't, because issues, problems, and consequences cross town lines, and the cumbersome nature of all this government costs too much and stretches citizen participation. Many of these entities would function better regionally than they do municipally. But people cling to the existing structure because it's the way it's always been, because they worry that the individual characteristics of the different towns would be lost with regionalization, and because of a concern about loss of control. Our lives are attached to the Vineyard, not just to a single town. If I pollute the water in my town, it will eventually have an effect on yours. The cars clogging our major intersections come from all six towns. We have difficulty dealing with our big issues because we fear losing town identity and control. From the point of view of subsidiarity, some of these activities would work better at a *larger* scale, a regional scale. Subsidiarity works both ways.

Although the point of localization is to take actions and create policies that discriminate in favor of the homespun, a local economy cannot exist in isolation. In our interdependent world, we must think in a variety of scales and plan regionally, nationally, and internationally, as well as locally. Can we reconcile these different scales? If so, the reconciliation will require a new set of economic attitudes, signals, incentives, and rules. Herman Daly, the economist and futurist, says that discussions of international trade rarely consider the huge energy and pollution costs of transporting stuff from place to place. He maintains that fully 50 percent of international trade is the simultaneous exchange of essentially the same goods. The Danes ship sugar cookies to America. Americans ship sugar cookies to Denmark.

"Why in the world," he asks, "don't they just exchange recipes?"

Local Enterprise and Ownership

To serve different purposes appropriately, we need big organizations and small organizations and everything in between. But current thinking is so committed to big that small gets left in the dust, even when it can do the job best. Take agriculture, for example. Today's agribusiness makes large quantities of food that must be shipped over long distances. A large percentage spoils before it ever reaches a shelf. In terms of freshness, nutritional value, and freedom from chemicals, the quality is low. The number of middlemen is large. Energy input is high. Fertilizers, preservatives, additives, and pesticides are pervasive. Small farms have declined and small-town life has eroded as agriculture has become increasingly corporate.

These quality factors have contributed to a lively resurgence of small-scale agriculture, especially organic agriculture, even in cities. Michael Shuman reports, "Over the past twenty years, New York City has opened a thousand community gardens on public land and eighteen public markets to sell produce grown in them." He goes on to say that, in general, the economics of small-scale food production is getting better: "If growing numbers of consumers continue to prefer food grown, raised, processed, packaged, and sold by local farmers, the economics of community food production will steadily improve."[8]

Communities are positively affected when companies like Stonyfield Farm create demand for high-quality organic agricultural products. Gary Hirshberg of Stonyfield says that his company, in building a market for organic yogurt, has caused more than one hundred dairy farms in New England to convert to organic farming methods by providing encouragement, assistance, and a steady market at good prices. These organic farms are thriving while traditional small farms are collapsing. The Farmers' Diner in Barre, Vermont, has a goal of buying all its food from farmers located within its watershed. It's well on its way to that goal, although nobody is growing coffee beans in the state of Vermont. The founder, Tod Murphy, aspires to make small "pods" of diners, four or five to serve a region, each pod served by a central commissary and processing plant, each limiting its food-buying radius. The centralization of agriculture into agribusiness is a recent phenomenon. Today, agriculture is beginning to rediscover the virtues of small-scale local production.

Energy production, likewise, is once again becoming more appropriate to small-scale production. As we (all too gradually) switch to conservation and renewables, and as hydrogen becomes a more likely prospect as a future energy source, the decentralized nature of new technologies has been responsible for a new industry of "energy service" companies (ESCOs) that began as conservation service providers and are now becoming small-scale producers. Michael Shuman again: "Judicious uses of land to produce food and solar energy to produce electricity demonstrate how community business can transform natural assets into a profitable product that serves local needs."[9]

As windpower has become more economical, new proposals are emerging in many locales, sparking controversy, dialogue, and, in many cases, acceptance. Europeans are leading the way, but as American communities find out from one another about the benefits, we will see more and more of this promising technology. An important benefit of wind is that it can coexist on land and water with other typical uses. It does not use up anything. It is cost-effective in windy areas today, and as it becomes more widespread it will become useful in less windy places. Wind energy is compatible with agriculture; the two endeavors can share the same land. If agriculture and energy move toward decentralization and localization, two of our most basic needs could gradually be met by local producers, leading us toward a healthier economy.

To make airplanes and steel we need large organizations, but most products, even high technology, do not require large enterprise. In the Emilia-Romagna region of northern Italy, small, locally owned firms involved in flexible manufacturing networks make an array of high-tech products. There are ninety thousand manufacturing companies in this highly prosperous region; 97 percent employ fewer than fifty employees. There are eighteen hundred cooperatives employing sixty thousand people. Emilia-Romagna is an inspiring example of partnerships among government, businesses, and cooperatives that has been working well for fifty years.[10]

Most businesses work well at a small scale, among them beauty salons and barbershops, banks, plumbers, auto repair shops, picture framers, locksmiths, health clubs, accountants, real estate offices, restaurants, lighting manufacturers, and doctor's offices. Small-scale, locally owned retailers can effectively use amalgamated purchasing for more economic power. It is possible to create, as communities, the enterprises and organizational structures we need.

Malta and Mad River

If you drive toward Malta, a crossroads town in northern Montana with a population of 2,300, you'll see the large grain elevators from miles away. In town, there are places to eat and sleep on Route 2. At the West Side Truck Stop the braying and stomping of the cows in the cattle trucks mixes with the sound of idling diesel motors. Main Street is several blocks long. The Village Theater is open from Friday to Tuesday. This is a typical small town of the American West, with wide streets, tall trees, and single-story bungalow-style homes. Billings, the largest city in Montana, is 200 miles to the south. There is little between except rolling prairie and the buttes and breaks, forested coulees, and desolate badlands near the Missouri River. Glasgow (population 3,250) is 68 miles to the east. Havre (population 9,800) is 88 miles to the west. Swift Current, Saskatchewan, is 142 miles to the north.

Until recently, a gold mine in Malta employed two hundred people. It closed in the year 2000. Not long after that, Anthony's, the local clothing store, closed its doors and you could no longer buy socks or underwear anywhere in town. It's a long drive to Kmart and Wal-Mart. Aside from the hardship of traveling that far for T-shirts, when people go that distance they also tend to take care of other needs—filling their prescriptions, shopping for groceries, and buying beer—thereby taking that business away from Malta.

Anne Boothe runs Malta's Chamber of Commerce and Economic Development Council. Vibrant, energetic, and optimistic, she's the kind of person you'd want leading your organization. She saw an opportunity, spoke to community leaders and businessmen, and inspired them to support the formation of a locally owned clothing store. They sold shares at $500 apiece, pitched their idea as an investment in the community rather than a path to wealth, and quickly raised $282,000. The store, Family Matters, opened in August 2000. There are 150 investors, many of them local businesspeople, and each $500 share entitles the investor to one vote. The five-person board and the investors are committed to the store—when it opened, teams of volunteers cleaned the store and stocked the shelves. The store is well lit, colorful, tidy, active, and well stocked with boots and blouses, sneakers and sweaters, Fruit of the Loom and Carhartt.

Family Matters has become an anchor downtown business and gives people one more reason to shop in town. According to Anne, the store has given the other businesses in town a boost. "It's not Wal-Mart prices," she says, "but it's not too bad. Our Carhartt and Levi's prices are pretty good."

Anne stresses the importance of the leadership that came from the existing business community. "If the downtown storekeepers look at it as competition, it won't survive."

She says Family Matters has a great manager and a strong board. "People are proud. It's a big community effort that brings people together and reminds us why we live here. It's about helping each other survive."

Toni Bishop, the competent, no-nonsense, fiercely dedicated manager, says, "I think the store will be here a long time. It keeps getting better.

Family Matters anchors Main Street in Malta, Montana. (Photo by author.)

The community has been tremendously supportive. I just wish the locals who are still set on driving to Wal-Mart would give us a chance. I see people in here with their kids and the kid sees something he wants and the parents say, 'Hold off, we're going to Billings this weekend.' Some people still don't get it."

But others do. The idea has caught on. Other community-owned clothing stores have opened nearby, in Plentywood and Glendive, both in Montana; in Powell and Warden, two towns in Wyoming; and in Ely, Nevada. Contrad and Colstrip, also in Montana, are working on it. More are in the planning stages, and the idea is spreading to the Midwest and the Northeast. Some of the stores are beginning to band together to do order splitting and collective buying.

The Montana clothing co-ops are owned by their communities. The owners made a different kind of investment than most: a financial investment in a community-owned business. It's similar to the kind of commitment we often make when we contribute time and money to civic activities in our town or city, but in this case the effort is to create a business that's essential to the community.

What makes these Montana communities so different from the many desolate, dispirited communities described in Thomas Frank's recent book, *What's the Matter with Kansas?* He says,

> Walk down the main street of just about any farm town in the state, and you know immediately [that] they're talking about . . . a civilization in the early stages of irreversible decay. Main Streets here are vacant, almost as a rule; their grandiose stone facades are crumbling and covered up with plywood—rotting plywood, usually, itself simply hung and abandoned fifteen years ago or whenever it was that Wal-Mart came to town.[11]

Is it simply the difference that one passionate person—or a small group—can make? Or is it something more fundamental—something embedded in different histories or different economic or cultural conditions? I don't know enough to make any informed judgment, but the question is interesting. One way or another, there's something special happening under the radar in places like Malta, Montana.

In the Mad River Valley of Vermont, Betsy Pratt, the owner of the beloved ski area Mad River Glen, was ready to retire and sell in the mid-1990s. Locals were worried that the character of the storied place would change under new ownership. Someone suggested a cooperative, so that the skiers and townspeople to whom the area means so much could maintain and operate it as they wished. A local group organized and began to sell shares for $1,500. They were unable to raise the full amount, so Pratt gave them an interest-free loan for five years, and the group bought the area in 1995. The debt has since been paid and the price of the shares has risen. No one can own more than four shares, and there is no profit sharing—the profits all go back into the area. It's another example of an investment in community. Mad River is an essential part of the valley, and the employees, townspeople, and users are at the helm.

The Mad River shareholders are committed to keeping for the ski area the qualities that compelled their allegiance. They've made an interesting corollary discovery: They have come to realize that their organization, which formed simply to maintain an important community institution, has the ability to tackle other issues important to their region. They have created a new framework for local problem solving and community building. Over time this new organization, this new locus of community power, is likely to be used in other creative ways.

The Mad River story and the Montana clothing co-ops demonstrate the difference between internal and external investment. Those who invest without a local interest in a business care mostly about financial return, whereas community members will often invest for nonmonetary return. Employee ownership, or in these cases community ownership, makes an enterprise nearly invulnerable to outside takeover. Why would the owners of these businesses, who bought them for local purposes, sell to someone else who has other interests? Michael Shuman says,

> If enough of us create our own corporations based on new visions of social responsibility, and if we choose to buy and invest only in these firms, other corporations will either adapt or die. If we create even a small number of self-reliant communities in which every resident has a decent job that produces basic necessities for one and all, other communities will visit, learn, and follow. We have far more power than we realize.[12]

Local business and industry need support, encouragement, incentives, and controls. Fortunately, despite whatever encroachments have been made by dispassionate big business (each new Wal-Mart ultimately puts an astonishing number of local businesses under), we still have our local economies. We don't need to take them back from global corporations; we already have them, in whatever condition they're in—good, fair, or poor. We can move them forward from here.

The Derry Restaurant

When Chris and I drive from the Vineyard to Vermont and back, we often stop for lunch or dinner in Derry, New Hampshire. Exit 4 off I-93 takes us to West Broadway, a road that, a mile to the east, becomes Broadway, the main street of Derry, a town of twenty-two thousand. Derry's H. P. Hood dairy farm used to supply milk to Boston. A major shoe manufacturing enterprise provided employment until 1960, when fire destroyed the plant. Today, these industries have been replaced by several high-tech manufacturing companies, while most of the commercial activity continues to consist mainly of small, family-owned business.

Around the I-93 interchange is a cluster of fast-food establishments and gas stations, transitioning to a mixture of houses and businesses as you head toward town. About halfway to the center of town is the Derry Restaurant. Whenever we go, the parking lot is nearly full. Inside, the clientele is mostly families. The menu is pizza plus varied home cooking. The paper place mats have advertisements for local businesses like Tires Too, Ben Franklin printing, and Jake's Transmission Clinic (whose motto is "We give a shift"). The lights above each table have fake stained-glass shades; the paneling is faux; the aquarium is large (kids are usually clamoring to look at the fish); and the servers are kind and friendly. In the bathroom, a sign says, "State law (and common decency) require that employees wash their hands before returning to work." There's nothing fancy or stylish here—just a warm, friendly, well-cared-for place full of common decency.

We have good meals there. When we leave, we pay the owner, Nick Samaras, who is stationed on the other side of a large horizontal opening near the entrance/exit. He makes the pizza, dispenses to-go orders, and takes the money. He also thanks people mostly by name. Those whom he doesn't know—like us—he asks about the quality of the meal. He's been there twenty-nine years, and most of his family works in the restaurant. He says that Derry is a good town, that he's never had any trouble here. Once I asked our server whether the restaurant was a good place to work. She said, "They are awesome to work for." She'd been there eight years. She could maybe make a bit more money elsewhere, but she said that the money was good here, too. Mostly she liked coming to work in a place where, she said, "I've never once been disrespected or belittled, and they're so sensitive to employees' needs. If you need a day or two off, or even a week, you never have to worry about losing your job." She added that Samaras family members do not feel entitled or privileged, like they're the owners and you're the employees. Everyone works together to do the job they're supposed to do.

So I say, "Spread the word." Make a good place to work and your employees will provide good service and make good food. Provide good service and food and people will come. If people come, life and prosperity will follow. Seems simple, doesn't it? It's interesting that *this* parking lot is always full, and the lot at the McDonald's down the road is not. There are relatively few travelers here (although Nick says that he consistently gets people like us who stop by chance and continue to come by intention); mainly Nick's place is full of townspeople.

The Derry Restaurant and the Farmers' Diner are down-to-earth restaurants in working-class towns where people go to have a decent meal. One is part of a dying breed; the other is a conscious new alternative. They both have many of the qualities of Carlton Supply, in Hartford, the subject of my friend Jamie's letter earlier in this chapter. The difference is that Carlton closed its doors. Derry is still open and the Farmers' Diner is just getting rolling. They represent the kinds of businesses we need to support. If we patronize local businesses when we have the opportunity, start more, support efforts to restore small agricultural capability, and develop new renewable energy sources, perhaps we can, in time, through localization, find more balance with the global economy. In Jeffrey Hollender's *What Matters Most*, he quotes Judy Wicks of the White Dog Café in Philadelphia, a treasured community institution that has been committed to local community and economic development for decades. He says that although she "applauds those who are working to tame globalization and reform the corporation," she is focused on "trying to engineer a more fundamental shift toward local economies as an alternative to corporate globalization. We're interested in a corporate model oriented around small businesses, small farmers, small buyers, small sellers."[13]

Some may see this as unrealistic. Maybe, but I'm with Judy Wicks. It's an admirable goal to try to restore commerce to a more local orientation, serving people more equitably, positively, and directly. While reform efforts must intensify to curb corporate excesses worldwide and to harness globalization, this remains a daunting task. There is much to be done in our own backyards.

At South Mountain we treasure the opportunity to limit our work to this complex little island that we've come to know well, that's still a fine place in which to live and work. Committing to the business of place—our fourth cornerstone principle—is an unconditional investment in the people and economy of a single locale. We've tossed our hat in the ring here and tied our future to the future of the Vineyard. We expect no windfalls, just a legacy of ups and downs, trials and errors, and rewarding collaborations. We're eager to see what's around the next bend and pleased to be able to take part in the evolution of this place. We're staying close to home.

Looking across the Edgartown Great Pond from the barrier beach. (Photo by author.)

Craftsmanship oozes from this meeting of stair tread, riser, and free-form tree post. (Photo by Randi Baird.)

⌈ 6 ⌉

CELEBRATING THE SPIRIT OF CRAFT

Clients for whom we were building on an eighty-acre property requested a single house large enough to accommodate their children and a growing cadre of grandchildren. The site we chose had beautiful old gnarly oaks that nobody wanted to lose. We proposed a scheme for two buildings rather than one. They would be connected by a roofed breezeway that stepped down and curved from one building to the other in a graceful descending arc, navigating the landscape in a way that allowed us to preserve the most important trees. At the same time, the two-building approach broke down the building mass into a scale more appropriate to the site. Our clients accepted the proposal, we completed design, and construction began.

When it came time to link the buildings, it became clear that our breezeway design wasn't right. Built as conceived, it would have upset the balance and harmony that our site designers had so carefully achieved. The design team met on-site to explore alternatives. We suggested many solutions to one another, but none resonated.

Then one of our designers, Derrill Bazzy, said, " It's clear what the right design is: no roof at all, just a path winding beneath the tree canopy."

"Of course," we responded. "That's it!" The clients, however, had specifically accepted our design with the condition that there would be a protected walkway between the two houses. Would they go for this new idea? It was my job to approach them.

I remembered being impressed, years ago, when a friend and I were visiting the Japanese woodworker George Nakashima at his studio in Bucks County, Pennsylvania. Nakashima took us for a tour of his shops. They

The path as it was meant to be, without a roof. (Photo by Randi Baird.)

were four small buildings separated by white gravel paths and Japanese gardens. It poured rain in sheets that day. At each building the doorways were set behind deep porch roofs, and on each porch was a bin of umbrellas. As we walked from building to building, we'd grab an umbrella, carry it with us, and deposit it outside the next. A clever system. When I met with our clients I told them the story. "So that's our design solution: you carry your roof with you, if needed, and at other times you'll walk under the trees and the stars." They immediately agreed, suggesting only that the umbrella stands be prominently located and beautiful to look at.

Design is looking for needles in haystacks. All too often, we stop before we have searched diligently for the ultimate solution. To be able to consistently bring an unrelenting approach to design, we need to overcome fear, the designer's toughest adversary. It's fear that keeps us, as our landscape designer Sanford Evans says, "from suspending judgment long enough to arrive at a good solution." Fear that we won't find a solution pushes us toward the easy or the early concept and makes us bail out too soon. We can never completely conquer the fear, perhaps, but experience helps us recognize it and move forward despite it. Embracing the uncertainty, we can stay with our apprehension long enough to find just the right needle in the haystack.

If we were not practicing the processes of designing and building as a single craft, we may not have solved the problem posed by that breezeway. The design for the walkway roof looked good on paper, but as we stood on the site and absorbed the space, we all had the same visceral reaction. The structure we had proposed was simply a violation. You had to be there, with posts ready to rise, to know it. And if not for the building crew, who were ready to dig in until they looked carefully at what they were about to do and alerted us to the problem, we may not have known until it was too late.

This chapter is about the craft of designing and building. Celebrating the spirit of this craft has become our fifth cornerstone.

Making Things

Our primary work at South Mountain is making things—like houses, breezeways, and pathways—that are visible, tangible, functional. Along with houses, we make many things that are parts of houses: additions, interiors, cabinets, furniture, lighting fixtures, landscapes, and even, sometimes, collections of houses—small neighborhoods. We try to make things—neighborhoods and houses and parts of houses—that will have lasting value for generations. It's surprising how few people in this country actually make things like this anymore, and how few make things at all.

Some people make things that are visible, tangible, and functional, but not durable, like TV dinners, hula hoops, and haircuts. Most people make things that are not things at all, like insurance, mortgages, financial transactions, diagnoses, therapies, and sales. According to the U.S. census of 2000, less than 10 percent of employed people work in production occupations. There are as many people in media, sports, and entertainment as in manufacturing. More people deliver and move things around than work in the construction trades. In this country, we just don't make much anymore.[1]

The shop— it's always humming. (Photo by Randi Baird.)

Manufacturing continues to flee the United States in pursuit of cheap labor. The Bureau of Labor Statistics' year 2000 chart of the thirty fastest-growing U.S. occupations is made up almost entirely of jobs in which people do things for other people. Home health aides, human services workers, and personal-care aides are the top three growth occupations. Physical therapists, paralegals, special education teachers, medical assistants, corrections officers, child-care workers, legal secretaries, manicurists, nursery workers, flight attendants, guards, insurance adjusters, and a host of other similar occupations fill out the list. Only the thirtieth-ranked growth occupation makes *anything,* and they are the pavers who surface the roads and parking lots needed to get all these people around and park their cars so they can deliver all those services to one another.

I'm glad, of course, that we have all these people to help, teach, advise, cure, entertain, fix our teeth, and do our taxes. But I think many of us suffer from not making things. Like Betsy in our office, who wants to be proud of the products made by her company, the quality of many people's lives—both workers' and consumers'—is diminished for want of authentic products made by authentic craftspeople.

Although not everything we do is about making things—we too give service, advice, and assistance—the making of things is central to our enterprise. To ensure the quality of our products, we have developed a process of integrated design, a way of assembling expertise and creating dialogue through which good solutions can be found and good value created. Everyone involved is important, from the client to the building designer, site designer, interior designer, carpenters, foremen, craftspeople, subcontractors, suppliers, bankers, building inspectors, town boards, and neighbors. All influence the design. An extended, intricate web of relationships creates each project. To conduct this process effectively, we need to oversee all aspects. Over time we have developed the skills to do it. On nearly all of our projects it is our responsibility to do development and planning, design, building, furniture and woodwork, the interior, and aftercare.

The Master Builder Approach

The integration of design and construction in the making of a building is known as the master builder approach. (A 1794 reference to this term in the *Oxford English Dictionary* says, "When a building is to be erected, the Model may be the contrivance of only one head.") Having the master builder as both designer and constructor was the predominant method of building in the Western world before the industrial revolution, but it is not how most buildings are made today. From the beginning Mitchell and I had an intuitive sense that the design/build approach was the right way for us. As we learned to build, we also learned to design. At that time in our lives, there was no separation between work and play. All activity had both in it. I didn't know what a vacation was. Vacation from what? Trips always had a purpose, but that didn't make them more work than play. It was just as much fun picking up a load of wood at the sawmill as it was stopping at the swimming hole along the way. Both were part of a process. Similarly, there was no separation between designing and building.

The master builder approach assigns responsibility for all aspects and trades to one entity, unlike today's conventional method, which fragments responsibility into a process that some have called the "relay-race" approach. The conventional method, a product of our overspecialized professional culture, begins with a client who describes a set of requirements. An architect makes a program of spaces and designs the building. The architect hands the design to engineers—who may not understand the design intent—to check for structural integrity, superimpose mechanical systems, and design the foundation. The completed drawings go out for bids to contractors who have been remote from the design process and may have even less understanding of the rationale. They may not agree that the building should be built as designed. Subcontractors, further disconnected and having their own predilections, bid on parts of the work. A landscape architect is employed to design site functions and equipment. Plantings and landscape construction are assigned to another contractor. The interior is passed to another set of people.

The rigidity built into the process by complex contractual arrangements and segregated responsibilities makes it difficult to make changes. When changes do get made they tend to cause unintended consequences, because the channels of communication and degrees of oversight are not always clear and effective. None of the various participants may have a large enough view to know that when a wall is moved to line it up with the structure below, which may have been laid out mistakenly, it may affect the lighting, the mechanical systems, the window placement, and the furniture layout. When the building is completed, nobody understands the whole well enough to show the occupants how to use it, care for it, and operate and maintain the building's systems. As the baton is continually passed, those who could fix problems when they arise (as they always do) are no longer there. When there are problems, each practitioner blames another or feels no particular responsibility. Squabbling and lawsuits are commonplace.

It's only in the past century that building has come to be practiced in this way. Some are trying to make relay-race building work better through greater collaboration, but in many cases the method remains contentious, problematic, and mechanical.

A master builder who guides and oversees the planning, design, and construction and has direct accountability to the owner can lead a more effective process. In our case this is not a single person but a whole company. South Mountain *is* the master builder, or, more accurately, the whole South Mountain team is a "master collaborator." As master builder, we hold the conviction that design and construction go hand in hand. We combine the theoretical aspects of design intent with the knowledge of what it takes to perform a successful construction job. This is a direct process—just South Mountain and the client, fully responsible to each another. Although many other important contributors are involved and become part of the team, this central relationship provides clarity and integrity to the complex process of making settings and buildings. As master builder, we are the protectors of the process and the product. We are responsible for the land, the neighborhood, the town, and the clients' interests. We are accountable for the needs and feelings of the many people who do the work.

The master builder system is predicated on the understanding that sites and buildings are layered; they develop over time and need care, adjustment, and continuous knowledge. The work is never done. The relay-race approach postulates that an architect will develop a miraculous concept, others will turn this wonderful vision into reality, and when it's done, it's over, and it sits on display like a sculpture for its lifetime, which is sometimes surprisingly short.

Bilbao

On the way to Mondragon, Spain, my friend Lee and I flew into Bilbao. We planned to visit the Guggenheim Museum, one of the most famous new buildings in the world. We landed at Bilbao's surprisingly beautiful and well-conceived airport. This is an airport you actually want to spend time in. We did. Everywhere we turned we found thoughtful craft, lovingly attentive detail, extraordinary daylighting, and playful drama. We were delighted with this modern complex designed by Spanish architect and engineer Santiago Calatravi. Suddenly we looked at each other, as if struck simultaneously with the notion that if this was the *airport*, then it must be true—Bilbao is a true architectural mecca.

When we were able to tear ourselves away, we picked up our car (in a multilevel parking garage set neatly into a hillside to reduce its mass and impact) and drove downtown, parking in a long, sloping lot beside a railroad track, in the shadow of Frank Gehry's striking, soaring Guggenheim Museum. We walked the asphalt, hiked up a long, poorly made stone stairway, and stood on a stark paved plaza at the museum's entrance. There was one lone stunted tree planted in the plaza. The signature titanium siding was sloppily installed and rusting in spots. We were shocked by our immediate impression of mediocrity. Inside was no different. The spaces were dramatic, but we couldn't quite figure out where the art was. High-quality materials were poorly joined and inappropriately assembled. Mystified, we wondered what went on here. How could such a monument be so disrespectfully made? Was it built by the lowest bidder? Were the workers who were charged with construction disdainful of the project or opposed to what they were being asked to do? Was this truly the realization of Gehry's grand vision? Or had the design been passed through so many hands that we were looking at a diluted, half-baked version of the original intent? Or was it Gehry himself who was purposefully flying in the face of the ideal of craftsmanship?

As we left, the bell rang on a cathedral a few blocks away. The sound was deep and resonant, issuing forth from a building created by a master builder hundreds of years ago. The cathedral had endured, and would continue to. In contrast, the Guggenheim appeared to be the temporary winner of a relay race, beating its chest in hollow victory, disconnected from the surrounding neighborhood and the city, a sentinel rather than a participant. Is this building like a hit song that spends six weeks atop the charts, becomes dated, and fades into obscurity? We had heard that this acclaimed museum had stimulated an economic revival in Bilbao, but this fact didn't change our disappointment. To us, the airport was a fine symbol of the city.

Perhaps there is a lesson here, however, about celebrating risky failures (I think of the Guggenheim as a massive failure). We need these failures. Without them, there would be little innovation. Sometimes craft is overly conservative. There are always new approaches to invent, and we do not want to be hamstrung by precedent. Breakthroughs keep coming in materials, energy, building systems, and aesthetic forms and insights. We hope to create a healthy balance between innovation and precedent.

If students in architecture schools were encouraged to take as much responsibility for serving neighborhoods and communities as for mastering studio critiques and their own ambitions, we might see better balance. The more stakeholders we gather, the more intelligence we focus, the more we will get buildings and communities that are not merely, as architect Bruce Coldham says, "settlements between the various designers and consultants, all defending their own turfs."

Essentials of Craft

Taking responsibility for all the elements of making buildings and landscapes—development and planning, design, building, furniture and woodwork, the interior, and aftercare—gives us a chance to craft good buildings and landscapes. Integration of practice promises that all the parts may work together appropriately. If they do, they will serve their purposes well and presumably will be kept, and enjoyed, for centuries. This is our hope for the places we make. At the core is craft, the critical unifying aspect of our work.

A desk with a single-plank, salvaged "sinker cypress" top. (Photo by Randi Baird.)

Craftspeople have strong feelings about that which they make and how they make it. Sometimes there are elaborate discussions in our shop about a single piece of wood—recognizing how the grain runs, how it grew, and where the strength is; considering the orientation that will work best for the function intended; speculating about how to tease out all its beauty and how it will finish. Conversation ranges seamlessly from the overall qualities of the product to the most subtle and detailed elements of the process. We have more disagreements about wood and joinery than about money.

Craftspeople take their time. You can't rush quality; it develops at its own pace. You can't rush the skills of craft, either. They have to be absorbed over time. I used to be a woodworker, but I didn't grow up with woodworking. I took it up with passion, but it came to me slowly and incompletely. My son Pinto, however, has been around it his whole life, and I've noticed that he has it through and through, in ways I never could. He practices it like a skater who has been on the ice since the age of four.

When we work as craftspeople, the pleasure of work can soar above the tedium that—at least some of the time—characterizes all work. When we make something well designed and executed, it is a telling of the truth. A thing well made reveals how we think and what we admire. Craftsmanship translates well to different scales—everything we make, from a knob to a neighborhood, can be imbued with craft. And it translates to different arenas as well; letters, contracts, drawings, phone calls, schedules, budgets, and relationships can all be as well crafted as a stool or a stairway, and for somebody, at least, each of these is worth embodying with craft. Work is a chance to develop worthy expressions of craft in all its parts. The spirit of craft leads us, like the Balinese say, "to do everything as well as we can."

When we need to hire new employees, we try to pick people who, inherent to their commitment to craft, enjoy a sense of playfulness and joy about what they do. Many people at South Mountain still ply their craft for pleasure, working on their own domestic projects in their free time. Craft trumps conventions about the separation of work and play.

The imperatives of craft create an internal set of standards devised by the maker (child development experts say that this sense of being in control is one of the most important characteristics of play). The craftsperson or team of craftspersons makes the whole thing, from start to finish.

The purpose of craft is generally not dramatic innovation but evolution and improvement, so standards are based on practice that refines the craftsmanship over time. It's not practice for something else or performance against an arbitrary or competitive standard. When we make a piece of furniture, there's a complete outcome, but it's practice because it's part of a continuum that is never complete. We are forever polishing our skills.

The superior work that results from creating a workplace that engenders the spirit of craft is what makes it possible for us to fulfill the wishes of those for whom our work is done. They are expecting something authentic to be made especially for them. But they are not our most difficult taskmasters. The difficult ones to please are the makers themselves. We, the makers, are never satisfied. Every design has flaws that we don't see until it's built. Each building could be detailed more coherently, each chair could fit the body just a tad better, and every color could be one shade closer to artful perfection. We grumble, we assess, and we curse our carelessness and foolishness. We are gratified when a tight fit leads to an elegant result. That is the craftsperson's lot.

I'm still heartbroken (that sounds extreme, but I can't think of a better word—it's more than sad, I know, so maybe it's *slightly* heartbroken) when I look at something we've done and see a room or a roofline or a detail that should have been better. I wish we'd come closer to the target. Often we put our heads together, in such situations, and think hard about how we could have, and whether it's too late. Sometimes it's not. Even when the solution takes serious reworking, we repair the cause of our slightly broken hearts when we can.

Panic happens, too, although less often than heartbreak, and it's usually unwarranted. There's almost always a way out from under our mistakes. For me such panic is partly a carryover from our early days on the Vineyard, before we understood the powerful abilities of wind-driven rain. We made buildings that regularly leaked whenever there was a storm. Leaks are loathsome. Nothing except fire harms a building more than water or diminishes, to a greater degree, the sense of shelter that buildings provide. When I would wake up in the middle of the night and hear the wind howling and the rains pounding, I knew the morning would bring phone calls and trouble. Over time we solved those problems through practice, observation, reworking, and repairing, but for many years the onset of stormy weather would provoke in me that sudden moment of alarm.

Hard to Find

It's not easy to find or train good craftspeople. The work is hard, great discipline is required, and the financial rewards are not as great as in many other sectors. The culture directs us elsewhere. Craft is no simple path.

There's an ad that airs on the the Weather Channel. Two snowboarders are standing on the side of a snowy road with their gear, trying to thumb a ride up to the mountain. "Yeah, dude, I think I'm gonna major in ceramics," says one. "It's so Zen. Just you and the wheel. I can see myself in ten years sitting in a little cabin, living off the grid, throwing pots in some beautiful setting." Just then a Range Rover stops and picks them up. They get in, settle into the plush seats, and gaze around, awestruck by the lush interior. They look at each other with raised eyebrows. The other kid says to the first one, "Well, you could minor in ceramics."

The Design/Build Process

The master builder approach has been our way of working from the start. The process has endured, but what we do with it has changed. At the beginning, thirty years ago, our geometries were simple, predetermined by our limited abilities. As our range has increased, we've learned how much more there is to know than we can know, and we continue to be excited by assimilating new knowledge and ideas that lead to the development of our own eclectic design language.

As we learn, our ambitions grow, too. There are qualities that we feel our buildings must have in order to be worth making. These are inspired partly by the Vineyard's natural environment, architectural heritage, and community and partly by our clients and our own aspirations. Well-crafted projects, in our view, embody a collection of qualities: they are artful, calm, comforting, light and airy, durable, flexible and easy to

Our former Chilmark house: crafted passive solar. (Photo by Jennifer Levy.)

change, energy efficient, healthy to live in, made from low-impact materials, surprising, uplifting, and, of course, satisfying to the needs and desires of our clients. All of this applies to both buildings and landscapes.

Most of all we wish to make buildings that are loved.

We keep these qualities front and center as we establish criteria and priorities, gather information, and synthesize solutions. When our ideas jibe with those of our clients and we all understand our roles, the result is an extraordinary shared adventure. A critical piece of the design journey is something we have discovered that may be self-evident to others: there is no better design tool than the act of experiencing a real building with a client. We are fortunate that much of our work takes place in a small geographical area. We've built and kept strong relationships with our clients and made a point of cultivating access. Once we've completed early questionnaires, design programs, and a rough budget, we begin to spend large chunks of time with new clients in and around past projects. Here is where we find connections with them, gauge reactions, and start to weave the fabric of the new building. This is what virtual reality and computer-generated walk-throughs are used for. They're good tools, but

I find that *actual* reality is even better. The conversations we have when we explore a space together with our new clients have more life and vitality, and more precision, than would be possible if we were only looking at drawings or models.

Memories return: *"The house I lived in that I loved best had windows just like this; I had forgotten all about that window pattern."*

Tastes change and develop: *"We saw these log posts in photos and thought they looked like Lincoln Logs, but seeing them up close they feel natural, expressive, and appealing."*

Requirements change: *"This is twenty-five hundred square feet? No way we need more than this; the thirty-five hundred we were imagining would feel like a mansion."*

We work in those houses together. We talk, we reminisce, we measure, we model. Frank expression is common:

"This room reminds me too much of my grandmother's house in Queens."

"If only the window tops and the ceiling were six inches higher."

"Can't stand granite—reminds me too much of Martha Stewart."

"Ugly fixture over the dining room table."

Honest reactions and heartfelt expressions give us the information we need. A pattern evolves that's full of individual meaning and will lead to something entirely new and different: their home, a place that's uniquely theirs and different from all others. While engaged in this exercise, we remind clients that when their project is complete, we'll want to use their house for the same process with future clients. No one has ever denied us this; people feel that it is only right that they extend to others the opportunity that was extended to them. We are grateful, because it's one of the best tools we have.

Our job, as designers, is to ask, listen, and suggest solutions. The clients' job is to articulate needs, preferences, and desires. Together, we test schemes and hypotheses to see whether they work for both parties. Conversations range widely, because design combines style, personal expression, technology, cognition, aesthetics, functionality, and the satisfaction of human needs.

Information from clients leads to dialogue. If a client tells us she likes houses with skylights in the bedrooms, we ask why. She may say that she likes morning light in her bedroom, or that as a kid she once stayed somewhere with a skylight over the bed and loved it, or that she always dreamed of having a skylight over her bed, or that she likes lighting from above. There may be conflicting information. For example, "I always dreamed of having a skylight over my bed" conflicts with "I like a cool bedroom in the summer," because skylights overhead will tend to overheat the space. The information that comes from asking the question "Why do you want skylights in your bedroom?" may in fact lead to skylights, or it may not. It may lead instead to dormers, or high clerestory windows, or a particular kind of artificial lighting, or skylights with shading devices.

We are each designers. Each of us makes countless design decisions every day—what to wear, how to do our hair, which ear to pierce and where to put the tattoo, how to arrange the dishes in the dishwasher, where to hang the picture on the wall, how to load the trunk of the car. Each design decision has a result that affects us. The ultimate result of a good design process is a well-mixed blend, a combined expression of clients and designers. To be good at being a client takes being good at communicating how one feels in a house or a landscape. To be good at being a designer takes good listening and asking skills and having the good luck to see our way to good solutions. When the process works well, it's hard to look back and say what parts and which ideas came from whom. At the heart of the design process is trust—trust that a good solution will emerge from the collaboration, if given sufficient time and space.

The relationship between the client and the designer/builder requires honesty, forthrightness, and the fulfillment of expectations. We must build trust from the beginning. If our clients are lavished with commitments fulfilled, they will come to expect it and trust that things will continue that way. They can loosen up and enjoy the ride. The pattern of honesty assures that they can be equally forthright with us. We grow to be allies and learn to work well together. We tell them when we feel that they're doing a good job, or not, and expect the same of them. When we screw up, we tell them we screwed up. Everything is on the table. This especially applies to the communication of bad news. It may be human nature to avoid bad news, but good work requires bluntness as much as kindness.

It's equally important that the chemistry be right. Experience has taught us that the best clients are people who are good at what they do, secure with their competence. These people are looking for someone who works the way they do. They won't pretend to be able to do our job; they know they can't, just like we couldn't do theirs. They are looking for expertise, integrity, reputation, and shared aesthetics. We are looking for people who appreciate our philosophy, our methods, and our work, and who want to learn, with us, what the result should be.

We deal forthrightly, and early on, with money. We talk about money from day one. It is an essential design constraint, and we must be able to discuss it as frankly and knowledgeably as siting, space, aesthetics, energy, dormers, and window seats. Because we have responsibility for the building as well as the design and a full set of construction skills and broad experience, we can competently advise clients about costs as we design. It doesn't matter how expansive or limited a client's financial capability is. Everyone wants good value, and everyone wants to know what they're paying for. We begin the design process with a program (how much space will be devoted to what), a concept (what is the scope of the project and what kind of project it will be), and a budget. The budget informs our thinking all the way. It's no use to complete a design that excites and satisfies if it will cost twice what the budget calls for.

Whether our client is one person, a couple, a family, a town, a committee, a tribe, or a nonprofit board, this process is key. Our design/build method makes room for many to be involved and, at its best, encourages a kind of "blooming, like a flower, and unfolding," to use the words of architectural theorist Christopher Alexander.[2] It accepts surprises and unplanned course changes.

In the movie *Little Big Man*, the Native American grandfather, played by Chief Dan George, one day determines that it is a good day to die. He makes his arrangements, calls for his buffalo robes, goes out onto the prairie, lies down, and waits. Time passes; it begins to rain. His grandson mournfully comes to check on him and discovers that Grandfather is still alive! How can this be? George rises with great dignity, gathers his belongings, and says, "Sometimes the magic happens, sometimes it doesn't." The chief is clear that his responsibility is to make a space for the magic, then to be ready for and accepting of the outcome, whatever it may turn out to be. This is also the key to the design/build process. All of our processes and tools are aligned, collected, and deployed for one purpose: to make room for the magic, in whatever form it comes, from wherever it comes, and whatever its meaning.

Standing there between the two houses, gazing at the landscape, struggling to design a roof for the walkway, we were all, however inadvertently, trying to make room for the magic.

"Aha," we are sometimes fortunate enough to say. "Got it."

Practices

Design magic does not come by chance. It must be invited. Many years of design/build work have produced a set of technical practices that are essential to our craft; the results we seek are predicated upon them. In each of our projects we embed four essential technical practices: (1) fully integrated site and landscape work done on a design/build basis; (2) specific tools and methods that promote durability so our buildings and landscapes age well; (3) extensive use of salvage, reclaimed, and carefully sourced lumber; and (4) emphasis on green building techniques. These practices have evolved over the years and apply to our most highly crafted, expensively detailed homes as well as to our most affordable housing.

Integrated Design/Build Site and Landscape

We create buildings, landscapes, settings, environments. We must learn to know the land, to recognize how our sites relate to the human and nonhuman communities around them, and to weld our buildings comfortably to the landscape. We do not have the skill to do that alone. Well over a decade ago, we realized that the quality of our landscape design and construction was not up to the quality of our buildings, so we teamed up with Indigo Farm, a landscape design and construction company. The people of Indigo help us achieve more satisfying results. Their participation often leads to unusual and complex approaches to siting, yet when they are finished, the sites feel like they have undergone little intervention. In the end, the houses feel calm and settled in naturalized landscapes of appropriate landforms and native plantings.

Sanford Evans, Indigo's founder, thinks like a sculptor. He sometimes talks about "transforms"—single transforms, double transforms, and triple transforms. It may sound like figure-skating combinations, but he's actually referring to the rotation of forms in the natural world. A single transform rotates a form once, in one plane. A double transform rotates the form in two planes, and a triple is a modification in space, like a mirror image.

We built a swimming pond for one of our projects. At the end of the pond was an ancient oak tree with a huge, reaching limb close to the ground. This limb had grown in a gentle curve to adapt to the complex pattern of stresses it had encountered. It had survived centuries of storms, snow, and growth. Sanford took the form of that limb, dropped it to the ground, rotated it 90 degrees, and embedded it in the curve of the dam. A single transform. It's not easy to recognize this simply by looking, but implied in the design solution is the acceptance of the story of the evolutionary success of that tree. This attention sets up resonances and imparts a subtle serenity to the landscape. As Sanford says, "I'm responding to the quality of the space and illuminating it. It's about loving where you are."

Our houses are similarly influenced. We might back a house gently into a hillside or step it up the hill rather than placing it in front of or on top of the hill. Branching patterns of the trees on a site are sometimes echoed in window muntins. The color of the house's shingles and window trim will match the silver-gray bark of the oaks and the light green of the moss that clings to them. Trees that must be cleared from the site are used to frame the porch or to hold up ceilings. The house (and the other "improvements" we make) is a foreign object in an existing landscape, so we attempt to gently celebrate what is offered and relate to the stories residing in the land.

We work with landscape designers from the beginning to the end of every project, from the initial conceptual siting to the last flowering plant. One of our early projects with Indigo illustrates what the collaboration can achieve.

New clients bought a property directly adjacent to a site on which we had built a house for other clients. The two sites shared an access. But when we looked at the siting for the new house, it became clear that using the established access would bring cars to the wrong place on the property, subject the house to the glare of headlights, and diminish the scale and privacy of *both* properties. Sanford saw that if we created a new access up a steep hill farther down the main road, we would enhance the site in several ways. To do so, however, would be complicated and expensive, and we doubted that our clients would approve the plan. It was too far-fetched. Nevertheless, we were certain of its value. When we presented the plan, our clients fully understood and endorsed it. The radical departure dramatically improved both properties.

Some years ago Indigo restructured and became an employee-owned cooperative much like ours. Both companies have matured together.

Buildings That Age Well

Once I took my mother-in-law, a surgeon, to tour some of our houses. As we headed home she said, "You know, I've spent my life working my fingers to the bone to help and save people. But all the work I've done will soon be gone. What you do is lasting. It remains here forever."

I wish she were right. I didn't have the heart to tell her that most houses built in this country may barely outlive their mortgages. Some of her patients will last longer.

The only thing we know for sure about buildings is that they change over time. Uses change, configurations are altered, finishes are updated. Parts wear out. Systems degrade and need replacement. We're beginning to think of our buildings the way landscapers think about landscapes. No landscape is ever completed when it is first created. It develops and matures over time. What if we made our houses the same way? Let's plan many possible ways to expand and alter our houses. Let's create forms that easily accept additions. Let's frame in the door to the future addition, bay window, and extra skylight, and let's plan for the built-ins. Let's imagine future solar collection and leave plenty of unobstructed south-facing roof area. Let's expect that extra dormer in the roof. Let's build certain walls so that they can easily be removed to open up spaces as needs evolve, and let's use engineered floor systems that don't need bearing walls. If we do this work well, we will enable these houses to develop as a landscape does. In a decade or two or five, house *and* landscape will be very different from what they were when they were first occupied.

Stewart Brand, author of *How Buildings Learn,* the seminal book about what happens to buildings after they are built, demonstrates that the only buildings that last are buildings that are loved. These are the buildings that are maintained and carefully readapted over time as different occupant needs develop. If we make buildings that are beautiful and functional *and* easy to maintain, operate, and change, they stand a better chance of being loved. We have now maintained, altered, and renovated some of our buildings for a quarter of a century. We learn from them every day. We keep a small crew busy repairing and doing small alterations and additions to the buildings we have built. They serve as an important feedback loop, telling us what's working and what's not.

Aside from using a variety of advanced design and construction approaches and low-maintenance materials and techniques, and along with actually *doing* the required maintenance, we have developed several unique tools to encourage better long-term building life. All new cars have owners' manuals. Why don't houses? Houses cost ten times as much and have a significantly longer life. Finishing a house is a bit like letting a dog off a leash; now comes the test to see whether she's truly ready to behave. But we know she won't behave without guidance. For a house, that guidance equates to operating and maintenance instructions for its owners. When you buy a house, its documented history should be a part of its contents. We provide every house with a detailed owners' manual that includes a listing of all project participants; recommendations for ongoing service relationships; building chronology; complete as-built specifications; design and program information; regulatory documentation; sewage disposal, well, and water-quality information; a history of all the materials contained in the house; operating and maintenance instructions for the site and building; and equipment manuals and warranties.

In addition, we've developed a second important tool that we call a "roughing book." It's a series of photos of all walls and ceilings, keyed to a set of plans, that was taken after everything was installed in the walls but before they were closed in. This gives us perpetual x-ray vision into the walls and ceilings. Our subcontractors have come to rely on the roughing books. Once a client returned home on a Sunday night, heard water dripping into the cellar from the wall above, and called the plumber. He arrived, took a quick look, and asked for the roughing book.

An owners' manual for a South Mountain house. (Photo by Randi Baird.)

He found the photo he wanted, which showed the location of the pipe he had figured was leaking, saw where there was a joint in the pipe within the wall, made a neat incision in the drywall, went right to the leak, and resoldered it. No fuss, no mess, no search-and-destroy mission.

These documents really begin to shine years later, when it's time to add a room, move a wall, or make a built-in. The longer a building endures, the more valuable these tools will become, as memories fade and more alterations and repairs become necessary. If I were a banker, I'd require an owners' manual and a roughing book for every building I financed. If I were an insurer, I'd do the same. If I were a realtor, they would make my job easier. If I wrote the building code, I'd put this requirement in it. If I were a buyer, I'd be pleased to find them on a shelf in the pantry, like hidden treasure in an attic.

The message of the owners' manual and roughing book is this: "We expect this building to last a long time. We want to assure that it does. We expect it to change during its lifetime. We wish to make it easy to change." To make it last and to make it easy to change, we have to make it easy to maintain and we have to communicate—to the people in the future who will work on it and live in it—what it is and how it was made.[3]

It says something else, too. It says, "We're proud of the way this building is made. We wish to hide nothing."

Salvage, Reclaimed, and Carefully Sourced Lumber
We have always sought to use materials that we believe to be worthy of our efforts. Wood is an essential ingredient of our craft, and for many years we used quantities of old-growth redwood, cedar, cypress, and Douglas fir. As we watched the quality of the material decline and the cost rise, we also became aware that we were using material that was disappearing at a rate that far exceeded its renewal. In response, we decided long ago to source wood in new ways. We made the use of salvaged and reclaimed lumber a priority, and our practices changed dramatically as a result.

We soon discovered that there are wonderful and widespread salvage resources, but that using reclaimed wood is an intricate and subtle undertaking. Successful use requires a good supply network, substantial inventory (because you can't just go to the local lumberyard and buy it), specialized equipment, and careful coordination of design processes to encourage best use. A saying goes, "Never try to teach a pig to sing; you'll waste time and annoy the pig." We learned how important it is that the particular material be appropriate to the specific use. The material cannot be shoehorned into places where it doesn't belong.

When we first began to use salvage consistently, there was some internal resistance within the company. Salvage required new skills, took a lot of work, made a lot of mess, and demanded extra care and dexterity. It seemed to cost a lot, too, because we had to inventory rather than order as needed, and because there appeared to be so much waste. Some of my partners worried about the buildup of a large inventory. But the practice gradually became institutionalized, the benefits became clear, and this endeavor soon became a special source of pride, a distinction felt by all of us. Today it's as hard for us to remember the presalvage days as it is to remember when "Made in Japan" meant cheap imitations.

In some ways the transition to salvage was a trip back to our roots. In the 1960s and early '70s we had salvaged from necessity. In 1972, for example, Chris and I were living in southern Vermont with some close

The yard outside our shop is a sea of salvaged wood from all over. (Photo by Randi Baird.)

friends, trying to buy a piece of land where we could live and work together. The following spring we managed to buy a scruffy piece of rocky woodlands on Belden Hill in Guilford.

We set up camp, started clearing land and making gardens, and began to build, with roughly $500 in our collective pockets and all the time in the world. We needed it. With hand tools (we had no power), a chain saw, and a couple of raunchy old trucks (a Diamond T flatbed and a Jeep pickup without doors), we went to work collecting material. We disassembled several barns, scrounged old windows, collected piles of roofing slate, and lugged home anything that resembled lumber. To make foundation piers we found a pile of old railroad ties that, having been replaced by new ones, had been tossed beside the tracks in Brattleboro. We asked the stationmaster if we could have them, and he said, "If you can haul 'em, you can have 'em." Easier said than done.

Late at night, when no trains would be coming, we drove the Diamond T onto the tracks. The ties were about a quarter mile from the access point. The going was rough, and we hadn't gone more than a few hundred feet before we were hopelessly stuck. By the time we got the truck out, it was nearly daylight. This wasn't working. The railroad tracks in Brattleboro are located on a narrow shelf between Main Street (about seventy feet above) and the Connecticut River (about fifty feet below). Both drops are precipitous. We tried to pull the ties up to a parking lot on Main Street by hooking onto them with ropes and towing with the truck. The steep, rocky bank foiled this attempt, knocking around the unwieldy ties until they broke the rope or slipped the knots. Finally, we went out on the tracks on foot and pushed the ties over the bluff and down to the river's edge. On the riverbank, we laboriously gathered them one by one and lashed them together into a big raft, and my friend and partner Smokey Fuller poled them down the river, like a Mississippi flatboat man, to a boat launching downstream, where the rest of us—Dennis McHone, Lou Botta, and I can't remember who else—waited with the faithful Diamond T. Smokey landed the raft successfully, and we finally had our foundation piers.

Four of us must have spent the better part of a week to get those thirty ties. They were worth only about $150, but that $150 was nearly a third of the budget for the whole house, and now we had what we needed for our foundation. That's the way it went. By early fall, the forty-eight- by sixteen-foot building was ready to move into. We had a coarse but livable wood-heated, kerosene-lit home. Years later, when a main house was built, that building became a shop. It still stands today.

These days our salvage wood comes from many sources, including wine and beer tanks, pickle and olive barrels, whiskey barrel racks, water towers, dismantled barns and warehouses, logging leftovers, driftwood, river bottoms. The sources of supply provide interesting stories and compelling histories that give the wood a new kind of life. Our clients are taken with these stories; they become a part of the soul of the houses. For example, our staple wood is reclaimed cypress that is mined from river bottoms. This material, known as "sinker" cypress, is timber

The Choctawhatchee River — there's serious treasure underwater. (Photo by the author.)

that sank to river bottoms in the South during the time, mostly around the turn of the past century, when the great old-growth cypress forests were logged. We deal with a man in the northern Florida panhandle. He salvages these logs from the bottom of the Choctawhatchee River, cruising more than a hundred miles of river—well up into Georgia—hunting for sunken bounty. He pulls them from the river, mills them into rough boards, and ships them to a kiln and mill, where they are dried and dressed to our specifications. We've been buying multiple trailer-loads of this reclaimed cypress each year for at least a decade. I met Adlee, our supplier, just once. I traveled to his home, saw his mill, and rode up and down the river on his funky homemade pontoon vessel, which has a deck-mounted winch that pops the logs off the river bottom after scuba divers locate and attach cables to them. Onshore, Adlee led me through the woods to an opening where we marveled at a remark-able twenty-five-hundred-year-old cypress, one of the few true giants still standing in the region.

The state of Georgia once impounded Adlee's equipment, claiming that the sinker logs belonged to the state. They turned out to be wrong. In the old days, the many small landowners along the river who owned the cypress swamps would fell the trees during the dry season. In the rainy season, when the rivers were high, the loggers would float the timber downstream to sawmills. First they would brand their logs individually, so they could be identified and sorted at the mills. It was like a cattle drive. Because the logs were so plentiful, those that got caught in eddies or vegetation weren't bothered with, and thousands eventually sank. Adlee and his crew prospect for these logs.

When he first started, Adlee learned the history, and over the years he systematically approached people in the area, asking them to dig out the old brands from the barn or the attic. He purchased and collected them. On the day of his court date, he entered the courtroom pushing a wheelbarrow full of sections of logs with old brands stamped on them. Then he produced the corresponding branding tools and showed them to the judge. He said, "Your Honor, I truly believe these logs belong to me. They've got my brands on 'em." The judge agreed and directed the state government to leave Adlee alone.

We use the superb wood that Adlee reclaims for exterior and interior trim, woodwork, cabinetry, and furniture. Its patina, warmth, and varied color are beloved by clients, as is the story of its origin.

Some of our best salvage teachers have been people we originally met through the North American Timber Framers' Guild, including Merle Adams, Jonathan Orpin, Jake Jacob, and Max Taubert, who specialize in sourcing and using this extraordinary resource in the resurgent timber-framing industry.[4]

We also try to use local woods that come from our area or region. The best of all is wood that is cleared directly from the site on which we're building. In our windswept and sandy coastal area this is tough, because mostly all we have is twisted, gnarly, undersize oak trees. We've begun to put these trees to use as naturally shaped posts and beams, with surprisingly pleasant results. Driftwood, too, has become a source of high-character railings, porch frames, and furniture.

When salvage or local wood is unavailable, we try to use lumber from certified sustainably managed forest. Our exterior walls are covered with white cedar shingles from trees grown on lands owned by Seven Islands Land Company, a family-owned business that owns more than a million acres of Maine woodlands. They have spent substantial time and money to have an independent evaluator, Scientific Certification Systems, investigate their operation, certify their commitment, and make recommendations for how they can move their program even further along the road to sustainability. The Maibec company in Quebec buys their cedar from Seven Islands. They make wood shingles. We buy their shingles.

Incrementally, our sources of supply become more refined and the percentage of well-sourced wood in our buildings increases. We use a hierarchy of criteria to make choices that ultimately get us what we want— fine material that suits both our craft and our environmental principles.

Green Building Techniques

Buildings have significant environmental impact. Resources are used for their construction, energy is used to obtain and process these resources and to operate the building, and habitat and natural landscapes are disrupted by their siting. A successful design/build process must be well informed about these impacts.

We use a wide variety of "green" strategies to help us minimize the environmental footprint of our buildings and landscapes. These strategies fall into three basic categories of environmental design and building solutions. The first of these is the bundle of standard South Mountain techniques that we incorporate as a matter of course, such as minimizing square footage and volume; renovating whenever possible instead of demolishing and rebuilding; finding uses for construction and demolition waste; siting buildings to minimize visual and habitat impact; providing for future solar use; protecting existing vegetation from construction damage; avoiding the use of pesticides, chemicals, and toxic materials; using native plant species in the landscape; placing glass and windows where they will bring in the most sunlight, daylight, and natural ventilation; achieving first-rate energy efficiency; maximizing water conservation; and employing salvage and locally produced materials, those with a high recycled content, and those that are easily recyclable at the end of their service life.

The second category includes strategies we offer at significantly greater cost to clients who wish to further reduce the environmental impact of their home: heat-recovery ventilation; passive solar heat; solar

hot water; state-of-the-art glass and superinsulation techniques; composting toilets; enhanced denitrifying waste disposal systems; and solar
and/or wind-powered electricity.

Finally, there are specific community-planning approaches that we try
to incorporate into development projects: attaching houses or clustering
them tightly; adjusting older buildings to new uses and today's standards;
creating shared systems and functions whenever possible; designing
pedestrian environments that isolate vehicles; limiting pavement and
maximizing green space; preserving and creating prime agricultural land;
and locating close to public transportation.

As my fellow owners pointed out when I suggested that we identify
ourselves as an "ecological" company, we are a long way from truly
achieving such a goal. There's plenty of talk these days about "sustainable" building and development. Sustainability has been defined in many
ways, but mostly as variations of this: "meeting the needs and aspirations of the present without compromising the ability to meet those of
the future."[5] I've always loved architect Carol Venolia's definition:
"Sustainability is when if you keep doing what you're doing you'll be
able to keep doing what you're doing." Nothing we do is sustainable.
Everything we do has impacts that affect the future. When we think
hard about the meaning of sustainability, it becomes apparent that our
practices are quite crude. So if we can't use the word *ecological*, which we
tend toward, we sure can't use *sustainable*, which is entirely unachiev-
able. But we are always trying to move closer to the elusive goal by
doing better tomorrow, with this essential part of our practice, than we
did today.

Our three categories of green building strategies originated with Marc
Rosenbaum, one of the most intelligent, knowledgeable, and inventive
systems engineers I know. On complex projects we commonly use his
consulting services. During our decades of collaboration with him, he
has been a tremendous teacher, constantly encouraging us to reach further and to extend our ecological endeavors. Rosenbaum and all the dedicated practitioners and thinkers we have come to know through
association with the Northeast Sustainable Energy Association—Bruce
Coldham, Jamie Wolf, Terry Brennan, Alex Wilson, Chuck Silver, and a
host of others—have been great teachers, and great friends.

Our local friend Kate Warner is working doggedly—with some success—to persuade Vineyarders to follow in the footsteps of another small island, Samso Island in Denmark, which supplies all of its own energy from renewable sources and even exports some. We support her efforts and hope to help her achieve this goal.

Tree post and rail in our current house. (Photo by Kevin Ireton.)

Modest Goals

Perhaps the most important and elusive lesson of the practice of our craft is that even with all our lofty goals and expectations, it turns out that our most successful projects are characterized by a sense of modesty. We are not wildly innovative (only mildly so, I'd say), nor are we drawn to the highly dramatic. I'm glad of this. I think it's important to know who you are and what you do well, and to stay focused on this.

We recently hired a new architect. About six months after his arrival I gave him and three other new employees a half-day tour of some of our houses. Afterward I asked Ryan what he thought. Was there variety, or was there a sense of sameness?

He said, "It was like checking out the produce at an awesome food store. Many varieties, all mouthwatering." Our houses are as different as broccoli and sweet potatoes, but they are connected by a thread of intention and an evolving practice of craft.

In the late 1600s the finest musical instruments originated from three families in the small Italian village of Cremona. First were the Amatis, and it is said that outside their shop hung a sign: "The best violins in all of Italy." Not to be outdone, their neighbors, the Guarnerius, hung a bolder sign: "The best violins in the world." At the end of the street was the workshop of Anton Stradivarius, and on its front door was a simple notice: "The best violins on the block." South Mountain is trying to make the best buildings on the block. Good buildings, well-loved buildings, not "important" buildings. Buildings that, first and foremost, serve the needs of the people who inhabit them by supporting and nurturing their health, satisfaction, productivity, and spirit. Buildings that spring from a strong sense of place and deep collaborations, that celebrate the spirit of craft, and that seamlessly arrange the elements of our craft.

Stewart Brand quotes the Duchess of Devonshire, speaking about a particularly unassuming room in her spectacular manor house, Chatsworth, which she has restored meticulously:

> Being in this room on a winter night, alone or with one or two great friends, the sparkling coal fire with its low brass-bound nursery fender, the familiar things all around, sitting on a chair

which becomes a nest with letters and papers and baskets and telephone scattered on the floor, dogs comfortably settled by the fire,
or near the draught of the door according to their thickness of
coat, is my idea of an evening happily spent.[6]

Says Stewart about this passage, "The distribution of the dogs—and her
perception of them—signals a room thoroughly grown into. Professional
designers have borrowed all manner of Chatsworth fabric patterns and
historical references and design inspiration, but they will never get the dog
part right."[7]

At South Mountain, we are struggling to get the dog part right.

Trying to get the dog part right. (Photo by Randi Baird.)

Looking out at Island Cohousing from the porch. (Photo by Kevin Ireton.)

ADVANCING PEOPLE CONSERVATION

In 1980 a woman in her sixties named Madeline Blakeley called me to ask if I would look at a piece of land with her. Her husband had recently died. She was a librarian, they had no children, and they had always lived in rented apartments. Her dream was to own a piece of property. She had $7,000 in cash. A realtor showed her a lot priced at exactly that, but all her friends had advised against buying it.

The steeply sloping lot was adjacent to the main road from Vineyard Haven to Edgartown. Traffic on the road was loud and visible. The property faced due south toward a beautiful little valley, a perfectly matched solar exposure and view. Except for the proximity to the road, it was a lovely site. I suggested that she could build an earth-bermed, partially underground house. "The southern orientation aims away from the road just enough, and the berming would dull the noise as long as the house doesn't open to that side. We can design the traffic right out of the picture." She was excited. Even though she didn't think she could afford to build, the idea that the land could eventually be sensibly used was appealing. She bought the property.

At about the same time, we were approached by Cathy Weiss, a single mother who was an old friend of ours. She owned a piece of property in West Tisbury and wondered whether we could build a house for her that she could afford. Her budget was too small, but we had heard that the Farmer's Home Administration had a rural housing program that provided very low-interest loans to low- and moderate-income people. I explored the program for both Cathy and Madeline and found that loans were available at the remarkable interest rate of 1 percent. We hoped to

build a nice passive-solar house for Cathy and an earth-integrated house
for Madeline, but the Farmer's Home Administration's fixed expenditure
cap did not take into account either the Vineyard's higher construction
costs or the long-term energy savings our houses would realize. We did
plans for simple, compact houses and submitted them to Farmer's Home
with a request that they raise the mortgage limit (from $40,000 to
$48,000) on each house because of the energy savings they would realize,
which we carefully analyzed and documented. After extensive bureau-
cratic wrangling that bounced between us, the local office, the statewide
office, and the national office (which, in the end, was surprisingly supportive), the increase was approved. Unfortunately, it still wasn't enough
to build what we had designed, unless we cut out all our overhead and
profit and reduced our labor rates to below cost. Additional subsidies
were needed.

Enter David and Pat Squire. The Squires had purchased a piece of land
in Edgartown and had a house designed by a Boston architect. They
asked if we would be interested in bidding on the construction. I told
them that we built only those projects that we also designed and that we
didn't bid on construction projects. They were disappointed, but they
persisted. Then it occurred to me: What if we gave the Squires a bid that
had an explicit "premium" built in, recognizing both that we did not typ-
ically do construction-only projects and, also, that we were in need of a
means of subsidizing the two Farmer's Home houses? I shared the idea
with the Squires, and they invited us to submit such a bid. Our bid, one
of three, was roughly $40,000 more than the next highest. They chose us
nonetheless, and we built their house. As it happened, this was the last
time we built a house not designed by us. We also built two splendid little
houses for Madeline and Cathy for $48,000 each by reducing our labor
costs, forgoing overhead and profit, and making up the shortfall with our
extra earnings from the Squires project. Dreams came true. Monthly
mortgage payments were under $200! The Squires became strong sup-
porters of affordable housing efforts on the Vineyard, and years later,
when the Island Affordable Housing Fund was activated to raise money
for affordable housing, David became an important board member.

It's a nice story, isn't it? This was the first and only time we ever inflated
the cost of a project to support affordable housing efforts. It was an

Cathy's sweet little passive-solar home financed by the Farmer's Home Administration. (Photo by Jennifer Levy.)

unusual situation. This experience gave me the idea that if we could be a reasonably profitable enterprise, we could devote a portion of our earnings to affordable housing efforts. This is what we've done ever since.

Preserving Community

The Vineyard's desirability is responsible for its serious affordable housing crisis. It wasn't so long ago that young people could find caretaking gigs, shacks in the woods, welcoming campgrounds, cheap land, and a host of housing options. Locals could build homes on the family back lot and stay here to raise families. Today, staggering increases in real estate prices and the high costs of island living have made it impossible for many to afford homes. With its housing resiliency gone, the community is endangered.

We are beginning to suffer significant outmigration by people who are priced out of the housing market. Others, who are fortunate enough to have bought cheap land a quarter century ago and managed to cobble together a house and a livelihood, find themselves with *excess* equity. They could not possibly afford the house they own if they had to buy it today. The future looks grim for their children. Suddenly, they find themselves the owners of real estate worth a million dollars. They could buy a major chunk of western North Carolina or northern Maine for that. People are bailing out. We're losing the essential fiber of our community, and the new buyers of Vineyard homes do not often fill the same important civic roles.

A community consists of a place and those who have a relationship with that place: the land and the people. Land conservation is a familiar concept, but now we have to think about people conservation as well. Affordable housing is less about houses or land or development and more about people who belong in and love a place being able to afford to stay in that place. Advancing people conservation has become our sixth cornerstone.

Neighbors and nannies, schoolteachers and social workers, truck drivers and technicians, artists and arborists, plumbers and plasterers, the town clown and the town drunk, whistleblowers and curmudgeons, peacemakers and troublemakers, politicians and taxpayers, grandmothers and grandfathers, sisters and brothers, those of different ages, abilities, incomes, colors, religious beliefs, sexual orientations—we need all these people. Each time any of them is forced to leave our community due to escalating real estate prices, the Vineyard becomes a lesser place. People conservation also relates to important jobs going unfilled, or being filled by those who are unqualified, or turning over quickly as people lose their housing patience and hope. It relates to a commuter workforce instead of one made up of familiar faces and old acquaintances. It's connected to stress, anger, child abuse, alcohol, and divorce. When we begin to lose continuity of generations, we lose an essential element of community character.

Tom Chase, the director of the southwestern Massachusetts office of the Nature Conservancy, says:

> Every time we lose a Vineyard family of long-standing, we lose a grey cell from our cultural memory. Gradually we forget the traditions of land stewardship. Affordable housing is [about] not only . . .

people caring for people, but also people caring for the land. Enough Islanders have to be around so knowledge of the land accumulates and transcends generations. Without restoration of our natural habitats and generations of stewards, we might as well live any-where.[1]

The story of affordable housing is a story about long-term community sustainability.

Sepiessa

Awareness of the Vineyard housing crisis has been acute for years, but fear prevented action. The conventional wisdom about affordable housing seemed inconsistent with the local population's sense of home. They imagined shabby projects and beat-up ghettos that were the antithesis of the familiar Vineyard mix of neighborhoods, villages, and rural landscapes. The conflict between the knowledge that something needed to be done and the assumption that the results would be unsatis-factory was ameliorated by a few model projects.

For many years our company has engaged in these affordable housing efforts, but it wasn't until the Sepiessa project that I could recognize noticeable community impact.

The Martha's Vineyard Land Bank was interested in acquiring the Sepiessa parcel, a large tract owned by the Nature Conservancy, in order to create the only public access to Great Pond in Tisbury. When the Conservancy agreed to sell the land to the Land Bank, one of the condi-tions was that a three-acre parcel (the minimum lot size in the town of West Tisbury) would be sold to the Dukes County Regional Housing Authority to use for affordable housing. The sale occurred, but for years the land sat unused.

In 1995 we proposed to the housing authority that South Mountain Company design, finance, and build a four-unit rental apartment com-plex on their Sepiessa property. Because the property was zoned for only one dwelling, we had to apply for zoning relief to increase the density under Chapter 40B, a Massachusetts statute known as the "anti-snob

Sepiessa is a four-unit rental complex with a decidedly homey feel. (Photo by Randi Baird.)

zoning law." The law was enacted in 1969 to allow developers of affordable housing that met specific criteria to bypass local zoning that
obstructed affordable housing or had the effect of making such projects
financially unfeasible. Chapter 40B provisions offer housing developers a
powerful stick. It has been used well in many cases and abused in others.
In our case, the town was fully supportive and this "friendly" 40B would
permit the project without forcing the town to make potentially undesir-
able and hard-to-enact zoning changes in order to accommodate it.

Our goal was to make a high-quality, small-scale rental housing complex that could serve as a model for other affordable housing projects. To
do so, we subsidized the project in three ways: by doing the design and
construction at cost, by asking some of our regular subcontractors and
suppliers to commit materials and labor at cost, and by using creative
financing (longer term than usual and at a below-market interest rate)
offered generously by the Martha's Vineyard Cooperative Bank.

Sepiessa has been used, over and over, as an illustration of what affordable housing can be. The most important aspect of this project, aside
from the tremendous spirit of collaboration that produced it, is that it has

caused many who see it to say, "That's affordable housing? You're kidding me. I can live with that. No problem!" That comfort level has helped spawn other efforts. Despite extremely tight budgets, we were able to create housing that has a feeling of hominess and quality. The design fits the site comfortably. It demonstrates that affordable housing can work for both its immediate beneficiaries and the community.

Subsidies

In runaway real estate markets, affordable housing needs deep subsidies. What little can be done without subsidies produces shoddy housing that communities find unacceptable and that occupants find to be distinctly unaffordable over time due to high maintenance and energy costs. We believe that subsidies are essential to projects that respect both their residents and the community at large. Affordable housing should be similar, in many respects, to luxury housing, except for being more compact, somewhat less detailed, and, most important, differently financed.

Historically, in the modern era, whenever we have been successful at housing the poor—and even the middle class—in this country, it's been due to subsidies. The G.I. Bill after World War II was the greatest affordable housing subsidy ever. It helped the families of our returning soldiers and partially fueled the postwar economy. It's interesting to note that by far the largest *current* housing subsidy in America—to the tune of roughly $80 billion a year—is the mortgage interest deduction. Each of us who owns a home with a mortgage deducts our mortgage interest from our taxes. This huge housing subsidy benefits the middle class (properly, I think) and also the wealthy (improperly, I think). It has no benefit, however, for the millions of low-income people who really need it. Subsidies must be appropriately directed.

In our community a wide range of people, and not just poor people, need subsidies. People making 140 to 150 percent of the median area income for our county[2] are shut out of the local real estate market. Cops, teachers, nurses, tradespeople, and even town managers need housing subsidies here.

Addressing this situation requires tremendous commitment and creativity, but this community has demonstrated such qualities in the past. If someone had said in 1984 that by 2004 the Vineyard would have a dozen new *public* beach accesses and a growing network of *public* trails across the island, no one would have believed it was possible. Everything was headed in the opposite direction. The Martha's Vineyard Land Bank, voted into existence in the late 1980s, has accomplished this and more. Private land conservation groups have also been tremendously effective. As a community, we have actively and generously funded land conservation of roadsides, vistas, beaches, farmland, and ecologically significant habitats. At present, roughly 20 percent of the island's land area is under some form of permanent protection. Now we are seeing that people conservation is equally important. Our business has been one of the many entities that have banded together to lead this new effort.

At present we concentrate on private fund-raising through the non-profit Island Affordable Housing Fund, which people in our company have helped organize and sustain. People are contributing in large numbers. Soon we expect to have a permanent public housing bank, like our land bank, to provide consistent long-term funding for affordable housing. We hope to be able to generate approximately $5 million a year in addition to the privately raised funds, which are now roughly $1 million a year.

Historian and Vineyarder David McCullough has been tremendously helpful to the Island Affordable Housing Fund's efforts. Talking to a group of potential donors recently, he said:

> We're failing here on Martha's Vineyard. We're failing in a more serious way than we know. What we came here for, what we love about the place, is eroding before our very eyes. The essence of civilization is continuity, and continuity must exist for everybody.
>
> It ought to become socially unacceptable among people of affluence on this island not to take part in helping to solve these problems. We ought to be saying to everyone, to ourselves, if you want to be here, you want to be a citizen here, you want to own a home here, you want to take part in the community here, open up your wallet and pay your part proportionately. Because if the people who need to live here year-round, who do the work, who make it work, can't live here, it's all going to collapse. Simple as

that. And this isn't charity. Let's forget that. This isn't charity. This
is reality. This is being members of a great community. And it's
emblematic of the oldest, simplest truth in the world: if you want
to be happy, do everything you can to make other people happy.[3]

Martha's Vineyard is a rarefied place, and it faces particular problems.
But housing issues are virtually universal in attractive places. Desirability
trumps affordability. There's a bumper sticker that says, "Houses:
Everyone Gets One Before Anyone Gets a Second." I can't imagine what
the Vineyard would be like—or what South Mountain Company would
be like, for that matter—if our housing market were governed by that
principle. But somewhere between the draconian egalitarianism of this
bumper sticker and the sticker shock of a real estate market run amok is
the balance that can keep the Vineyard community healthy.

Employee Homes

As our company works to help others, our employees must have stable
housing, too. They've come to it in a variety of ways, mostly through
their own inventiveness, diligence, family help, and hard work. It has
been a primary South Mountain objective to help our employees satisfy
their housing needs, and we have found several ways to pitch in. We provide discounted materials, salvage materials from renovation projects,
and work parties to help with shingling, landscaping, and other tasks that
large groups of people can do together. The company offers employees
who are first-time homebuyers a cash stipend to help with down payments or necessary repairs.

We also extend flexibility and patience to our employee owner-
builders. The majority of our employees have built their homes from
scratch or done major work on fixer-uppers, often while they worked
full time. When you're building your own house, your evenings, week-
ends, and vacations are chock full of house building for a year or more.
It takes its toll. You're exhausted at work. But there's a big payoff at the
other end. Tremendous learning occurs when the responsibility for what
you are building is all yours. When they are done with working on their

Derrill and Joanne's house. The tree posts have become a recurring South Mountain design element.
(Photo by Derril Bazzy.)

own houses, our employees return more skilled, knowledgeable, and competent.

Of far more impact than the occasional South Mountain–organized work party is the informal help and exchange that goes on between individuals. A great bartering spirit connects these houses. Some people help others for years, and help subcontractors with their houses, too, and get paid back when their turn comes around. Those who put in significant hours often get paid on a moonlighting basis.

It's rewarding to see people find ways to provide housing for their families, but the fun part is seeing what they make. These are people who work every day, year after year, making homes for others. While they're at it, they form wide-ranging opinions about what they like and don't like about the homes we build, about what works, about what *they* will do when they finally get an opportunity to build their own home. When the chance comes, they're ready, and what they make can be surprising.

The houses built by our owners and employees are all very different and have tremendous character. Phil Forest and his wife Wendy designed and built a house that goes beyond our usual practices in terms of green building and the use of renewable energy. My son Pinto and his family are completing a house that is beautifully designed, fanatically crafted, and rigorously green. It's lovingly made in the same way he makes a piece of furniture in the shop (and it goes far beyond his parents' house in terms of detail and craft). There are many other examples. And on a regular basis employee owner-builders develop designs, approaches, and techniques that inform and change the company's work. Many years ago Peter Rodegast built a house for himself and used old slate roofing tiles to make a beautiful (and inexpensive) floor; ever since, old slate roofing tiles have been one of our favored floor and counter materials. Derrill Bazzy and his wife Joanne were the first, I think, to use the gnarly native oak posts that have become a common design element of our houses.

There's plenty to admire and much to learn from this illustrious group of homes. Our employees contribute to our enterprise in many ways while they create housing for themselves that assures a stable workforce for the company.

Development

We were in a Chinese restaurant with Chris's mother. She cracked open her fortune cookie and found a blank piece of paper. She asked the waiter whether he'd ever seen anything like that. He nodded gravely and said, "Umm . . . your fortune—no news is good news." That's the way many people feel about developers: no news is good news. Real estate developers are widely disrespected and mistrusted, but *development* need not be an ugly word. Development is invention.

Our affordable housing efforts have led us to development work. At South Mountain our first development question is, are we proposing to invent something that the community needs? If not, why bother? Let's design something else. When we examine community needs of our small island, we find the need for affordable housing nested in a constellation of other needs:

- open space preserved in perpetuity
- restoration of agricultural land

- places for low-impact businesses that will not contribute to strip development
- community systems for converting waste to nutrients, to protect our sole-source aquifer
- neighborhoods that encourage social interaction

So . . . let's invent all that.

My belief that we can do so stems in part from my tour of Danish cohousing neighborhoods in 1990. The cohousing concept, which originated in Denmark in the late 1960s and 1970s and has since spread to other countries, takes many different forms, but there are five general principles that set it apart from traditional residential developments:

1. The development is limited in size to twelve to thirty-five homes (large enough to avoid being a fishbowl and small enough that residents can all know one another).
2. The automobile is relegated to the perimeter, leaving pedestrian space within.
3. Houses are tightly clustered or attached to promote social contact and leave open space undisturbed.
4. A "common house" is included where residents often share one or more meals a week, where guests can stay, and where a variety of activities take place. Not only a community hub, the common house provides space not needed on a daily basis, thereby allowing individual homes to be smaller.
5. The residents are the developers, making decisions as a group and, in the process, creating community bonds.[4]

Island Cohousing

One evening, years ago, during a talk about Vineyard housing, I floated the idea of a cohousing neighborhood on the Vineyard. Two couples approached me and said, "This is just what we need here. Why aren't we doing it?"

I replied, "I've just been waiting for a few people like you. Let's go."

Those two couples, Randi Baird and Philippe Jordi and Sylvie and Paul Farrington, became the energy and the glue behind the beginning of Island Cohousing, which grew into a broad partnership among our company, sympathetic bankers, flexible regulatory officials, and a group of dedicated people who shared a dream and wished to develop and live in this kind of neighborhood. South Mountain was hired to conduct a land search, facilitate the formation of the group, and develop, design, and build a deliberate neighborhood. Through trial and error, dialogue, and persistence, we invented, together, a development that combines the elements I've mentioned. Its commitment to affordability did not diminish its commitment to be as green, beautiful, and democratically inspired as we could make it.

Our job was to provide the group with reliable information and guidance. This took us beyond the traditional roles of developer, architect, and builder. We also had to teach the language of development and the skills of effective decision making. We helped the group articulate values and aspirations, understand limitations and constraints, and formulate a sensible, real-world expression of its desire.

Early on we decided to do a combined project that would include both the cohousing neighborhood and new facilities for South Mountain. The timing coincided with the company's need for more space and a larger property. We were inclined to pursue the synergy of colocating our business with this innovative development. There would be major benefits to both organizations. South Mountain would share land and infrastructure costs to reduce Island Cohousing's costs, and Island Cohousing would be able to provide the business with the space it needed. We were able to acquire a large piece of wooded land that satisfied all needs.

Taking full responsibility for this project was a stretch for South Mountain. It was a larger endeavor with greater complexities than any we had previously undertaken. It included two separate but closely linked major elements: the South Mountain complex and sixteen Island Cohousing houses and facilities. We served two separate entities with fundamentally different needs: ourselves and the Island Cohousing group—two tough clients. The cost constraints were stringent; there was no

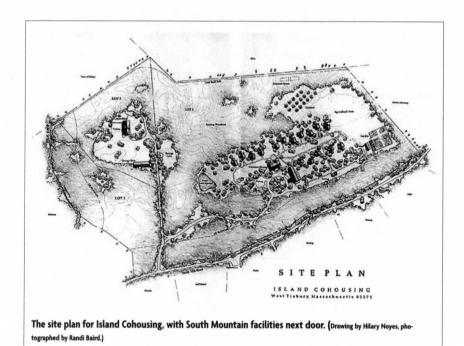

The site plan for Island Cohousing, with South Mountain facilities next door. (Drawing by Hilary Noyes, photographed by Randi Baird.)

wealthy client to absorb potential overruns. The regulatory hurdles were legion, and the potential for delays and conflict was real. It was a risky, difficult project.

The goal was to make a beautiful community that was affordable for year-round working people, in an area with very restrictive zoning. The project violated the then-current zoning in eleven different ways. I consulted a number of friends in local politics. They all said, "Yeah, nice idea. It makes a lot of sense. You'll never be able to do it here."

Then I spoke to our banker, Rich Leonard, president of the Co-op Bank. I figured if anybody would finally douse the flames, it would be him. Rich, who was one of the heroes of Sepiessa and other Vineyard affordable housing efforts, is a local boy who grew up in Oak Bluffs. He sat back and quietly listened to my urgent pitch. At the end he said, "Sounds like the neighborhood I grew up in. We don't have many neighborhoods like that anymore. There's an obvious need, and it's the kind of thing that could have a big impact in this community. Let's do it." Rich was a strong ally and supporter from the word *go*. His bank assembled the construction financing, committed to end loans for all the buyers, and provided discounted mortgages to the subsidized houses we set aside for low- and moderate-income earners.

The unknowns of the project prompted concerns within South Mountain. At a 1998 board meeting, when we were well into the planning, someone asked, "Is there any way we could lose our shirt doing this project?" This provoked a lengthy discussion. Someone suggested a "shared risk" arrangement with the cohousing owners in case the project went over budget, but the cohousing owners were already pushed to the limits of their financial capabilities. We would have to minimize the risk through good design choices and careful planning. In a memo to the South Mountain board after the 1998 meeting, I wrote:

> The tension here is internal: how can we do something we will be proud of and do it within the budget? The challenge is ours alone. [We] hope the project will at least break even . . . but there are no guarantees. One of the many variables is that it's most important that we are proud of what we do—if we're not, we'll have failed in a much bigger way than financially. We need to consider how this project affects the company's future. The whole point is to learn to do good housing affordably and to make that a part of our staple diet. It's part of a lengthy process that we've been working on for a decade and will surely be working on for at least another.

It was up to us to figure out how to succeed. The cohousing households were already sharing the financial burdens by paying portions of their down payment to finance the land purchase and planning costs, before they even knew whether the project would go forward. Their funds were not at risk because the land value secured them, but their money was tied up in something that could fall flat. It was up to us to shoulder some risk, too.

The regulatory process promised to be difficult. Once again we had to employ the state law for affordable housing, Chapter 40B, but compared to Sepiessa, this project, because of its size and combination of elements, would have far greater community impact and would receive intense scrutiny. It included the location of a business in a residential area, something that had never been done before in a 40B project. As well as town approval, it needed approval from the Martha's Vineyard Commission (MVC), a regional planning agency with broad regulatory powers. The commission determines whether proposed development projects over a certain size (which are known as "developments of regional impact") will be more beneficial than detrimental to the community. Developers often see the MVC as a difficult adversary. Because our purpose was fundamentally aligned with that of the commission—to shape a better community—we were able to work collaboratively to achieve an even better end result. When the commissioners finally voted approval, they issued a decision with fifteen conditions. All of these began with the words, "We accept the applicant's offer to . . ." We were pleased that our project was able to answer their questions.

Navigating through regulatory mazes can be trying, and given our track record in this regard, people with difficult land-use regulatory problems often come to us. We explain that we have no magic; what we do is straightforward. We will not bring a project into the public arena unless we believe it is good for both the land and the community. Others may not agree with our assessment, but at least we can bring proposals forward with full confidence that we have considered their consistency with our expressed values.

Once we had steered the cohousing project through the regulatory maze, there remained a host of complex financial, design, construction, and social issues to work through. Aspirations were high. All members of the group wished for first-rate ecological land use and building quality. They wished for low prices. They wished for community diversity. They wished to accommodate those less fortunate. We had to learn together how difficult and expensive it is to achieve all this, how tricky the balances are, and how substantive are the sacrifices that must be made.

Nearly half the houses were subsidized in two ways:

- light subsidies achieved by internal price structuring that shifted a higher percentage of the shared costs (development and design costs, infrastructure, and common facilities) to the larger houses
- deep subsidies from a combination of cash fund-raising from private donors and reduced mortgage rates from our two public-spirited banks

The first method kept the price low for smaller houses; the second allowed four of the houses to be sold to qualified buyers (those who made less than 80 percent of median county income) at a price they could afford. In addition, one house is rented at affordable year-round rates. The four deeply subsidized houses required that we raise $300,000 in cash. Donations were channeled through the nonprofit Island Affordable Housing Fund. Most of the money was raised from our past clients, who responded magnificently. Their contributions, and others, coupled with a foundation grant and a major contribution from the Martha's Vineyard Co-op Bank in the form of discounted interest rates, pushed us to our goal.

Along with the internal and external subsidies, there were four other significant cost-control mechanisms: production building methods and customization control, carving off and selling several building lots to offset land costs, reduced rates for South Mountain's design and construction services, and infrastructure sharing. The toughest was the first: working with a diverse group to maintain cost-effective design and production building. This was achieved by designing for flexibility in such a way that we avoided building sixteen custom homes. Customization has often been a virus that infects cohousing projects, because the members of the group making the overall design decisions each harbor different individual needs and desires. It's impossible at once to achieve low cost, high quality, diversity, and custom homes for all. If this is not well understood, it is likely that costs will spin out of control.

We had the good fortune to benefit from the experiences of others who had been through this process. We toured other cohousing projects and queried extensively. We listened carefully and took the advice of those who had covered similar ground. Ultimately, we limited customization while making room for individual needs and desires. We developed a single core plan that had a fixed number of expansion options. The house design is a basic twenty-two- by twenty-six-foot two-story rectangle with a full cellar. The public areas are on the first level; two bedrooms and a bath are on the second. Options included a third bedroom and a second bath, a fourth bedroom, and several bump-outs. The smaller houses are designed to easily allow the additions later. There were several landscape options, several exterior finish options, and

The interiors are warm and open. (Photo by Kevin Ireton.)

approximately twenty interior finish options. A one-page menu determined what each house would be, and a master chart tracked the options
in all houses. The group's self-discipline enabled agreement on a tremendous number of choices, right down to the tile selections and interior
paint colors.

The strong environmental aspirations for the project clashed with our
cost constraints. Choices were necessary. We told the group that the stan-

dard and not-so-standard green building approaches that South Mountain regularly uses (including finishing our houses—inside and out—with salvage and certified lumber, first-rate energy efficiency, and extensive use of recycled materials) would have only minor cost implications. Other possibilities would have major cost implications. We discussed generating our power with wind and photovoltaics, installing a district heating system fueled by wood that grows each year on our land, saving and protecting the native vegetation instead of mowing it down for construction convenience, converting our human waste into valuable nutrients by using composting toilets, and more. Some measures might be possible in the future; some would not.

The group decided to put its limited resources into two important and costly measures that could be done only now, and not later. The first was mapping, protecting, and saving hundreds of key trees and areas of native vegetation on the wooded site. To facilitate construction of dense housing, most sites are fully cleared before construction begins. After completion, trees and vegetation are planted. For many years after, the sites look raw and young. Protecting and working around existing vegetation is difficult and costly. We figured that this measure would add approximately $75,000 to the cost of the project. But what kind of landscape would that money have bought, spread over the four-acre disturbed area? Very little. Just a year after completion, once the plantings and lawns around the houses had grown in and the buildings had begun to weather, the neighborhood felt like it had been there for ages.

The second measure was to equip the neighborhood with composting toilets—a big expense, a lifestyle change, and a risk. Many in the group were strong supporters of this measure because it did two things at once: made waste into useful nutrients and protected the island's sole-source aquifer. But given the unusual nature of composting toilets, would people want these houses, would banks finance them, would appraisers value them, would the town allow them? These questions turned out to have positive answers, and this is one of the few new neighborhoods in the country (or perhaps the only one) that has no flush toilets.[5]

Perceptions and Results

During the design process there was a gradual shift in concern from individual interests to community interests. Members came to understand that the basic shape of a neighborhood is the fundamental layer that is hardest to alter, whereas they could easily change and personalize their own homes and adjacent landscapes over time. A transformative moment in this process came when we considered the exteriors of the houses.

South Mountain provided several exterior options that, along with the different sizes and configurations of the buildings, would mitigate against uniformity, including different porch roofs, a selection of roof and window colors, and optional exterior detailing. When considering these, someone said, "There's no way I could ever have that dark green color on my windows."

Someone else said, "Hmm . . . to tell you the truth, I don't care that much what I have on my house. I'm more concerned about what you have on your house—that's what I'll be looking at."

Aha! The group quickly agreed that we, as the designers, should survey members and take personal preferences into account but, of greater importance, that we should plan a pattern that would feel intentional and well designed when considered as a whole. At that moment the balance between group and individual interests became clear, and the ethos of group decision making became robust.

Perceptions and desires changed dramatically, too. Our most useful model was a beloved neighborhood in Oak Bluffs that began in 1835 as a Methodist campground and meeting space. Gradually the canvas tents, which were organized around a central tabernacle, had been replaced with small Victorian houses, and it had become a splendid little village. There are roughly three hundred small houses in the neighborhood today, and most of its traditions and architecture are intact.[6]

We put the "campground," as it is commonly known, to good use. We spent time there together modeling and measuring to help us determine our community layout. We stood on porches across from each other, talked, and listened to how it sounded. We measured the spaces between houses and discussed what different spacings felt like. I remember walking back to the car with one of the group members after our first campground

The trees were saved; the neighborhood feels like it's been there a long time. (Photo by Kevin Ireton.)

session. She said, "It's lucky our property is so large because I would never want to live in a house less than a hundred feet from my nearest neighbor."

I felt a momentary sense of defeat. This was not where we were supposed to be heading. Yet three years later she was content—no, excited—to move into her new house, which was twenty feet from the nearest neighbor. She had come to understand that fashioning a true neighborhood with proximity and intimacy was preferable to the suburban development pattern that scatters roads and houses throughout the landscape, in little relationship to one another. Less than 20 percent of the cohousing property is developed, and the resulting neighborhood captures, in its own small way, some of the endearing qualities that make the Oak Bluffs campground so renowned.

Architecturally, Island Cohousing is less about the houses and more about the spaces between them. In lieu of a street, a pedestrian common separates the buildings. Extensive common facilities—the common house, a swimming and skating pond, a community garden, woods and trails, and a basketball court and playground—favor neighborliness and community interaction. But such design elements can only facilitate what in the end is a tricky balance. Neighborliness cannot be mandated. Pushed too far, the notion of community can become forced, phony, and contentious. The diversity in incomes, backgrounds, and values of cohousing residents can lead to controversy, stress, uneven levels of participation, and occasional bouts of what Paul Farrington calls the "dysfunc-shui"[7] of neighborhood. Working out differences is not without conflict. There are a few people who don't much care for one another, and this will probably always be the case.

We have succeeded, to some degree, in creating a neighborhood that enjoys some built-in provisions that make it work differently than most. The neighborhood has a special sense of community. There are more connections among the residents, the connections are less random, and there is a splendid, convenient environment for kids, who can just walk outside or next door and find their playmates in an unusually safe setting.

It's instructive to watch our developing ability to work together, think together, and live together in a neighborhood. There are issues with kids, parking, pets, participation, consideration for others—all that one would expect. There is also a reserve of good feeling, which ebbs and flows, but which holds us together. Gradually, we have learned one of the essentials of consensus decision making: the art of gentle persuasion and compromise. We continue to deepen our appreciation of the cultural shift embodied in the concept and practice of cohousing and see in this shift much that is hopeful for the future of our neighborhood, our community, and our island.

Chris and I had not, originally, planned to live in the cohousing development. As part of the original development plan, South Mountain had purchased a six-acre parcel from the original fifty acres, and Chris and I had purchased another. These sales helped bring down land costs for Island Cohousing. We were planning to build on our own nearby property. But during the design process we began to feel a tug. The urge to "test drive" this new creation became increasingly strong. One night, as we deliberated, my daughter Sophie, who was fourteen at the time, said, "Now wait a minute, let me get this straight. You're working yourself to the bone making this wonderful community, and we're thinking of building a house down the road on the adjoining property? Hmm . . . seems pretty sketchy to me." This was precisely what we had been

The Common House: spaces we don't need every day. (Photo by Kevin Ireton.)

coming to feel. Soon after that we decided to become members, buy a house there, and move in. It's been a wonderful place to live—full of great moments, great failures, great frustrations, and great progress.

South Mountain Company and Island Cohousing

The four-year process of conceiving, developing, designing, and building Island Cohousing was an intense learning experience. The mistakes we made and the lessons we learned could fill another chapter. Living in a cohousing community is a continuous process of learning. Would we do it again? We would, and we will. This was a first-rate collaboration, with tremendously satisfying results. Island Cohousing was a defining project for South Mountain Company, and it has become a wonderful home for many islanders. The design of the community works. The design of the houses works. The group that developed the community plan worked together remarkably well under sometimes trying circumstances. Many organizations and individuals collaborated at a high level. The collective efforts developed the foundation of a vital, evolving neighborhood, one that we hope will encourage people to interact humanely with one another and with the landscape. We expect it to improve over time, both physically and socially. We see it happening already. Finally, Island Cohousing has been a tremendously effective model. It has received many visitors, some of whom confess that the *idea* seemed odd and insular, but when they *see* it they often come away inspired.

Ultimately, the creation of Island Cohousing served as a partial catalyst for surprising regulatory reform. The town planning board looked at this project, at Sepiessa, and at others and said, "These are exactly the kinds of projects we mean to encourage. Yet they violate our zoning in so many ways." This discussion was one of the factors that evolved into a process of rewriting the zoning regulations. The voters approved the new zoning bylaws, and today we could propose either Island Cohousing or Sepiessa without violating the regulations in any way. Small stones can make big ripples. Models are expressive and can be the antidote to fear of the unknown.

Jenney Lane

Just after Island Cohousing was completed, I gave a talk about affordable housing at a local church. An older couple, Ralph and Olivia Jenney, came up to me afterward and said they'd like to meet. We did. They had land in Edgartown, and they wanted to do some estate planning that included committing a part of their land to affordable housing. It turned out to be an extraordinary piece of land.

The land is in downtown Edgartown, five acres hidden in the middle of a well-developed old neighborhood of small homes. It's an odd-shaped parcel, patched together over time with tentacles that reach out to several surrounding streets. When you travel down the narrow dirt driveway to the Jenney house, you enter an enclosure surrounded by dense vegetation. It feels like you're away on an old farmstead, rather than in the middle of town. Ralph and Olivia wished for half the land to remain in their family, with the rest devoted to an affordable housing development right in their own backyard. This is one of those rare examples of someone saying, "Yes, in my backyard." I was impressed.

Ralph and Olivia were willing to sell the land at a bargain price to the

Island Affordable Housing Fund. We negotiated an agreement and began the preliminary design process. The opportunities were tremendous. The walk-to-everywhere potential of the in-town site became an immediate driving force. It was clear that at this location, if possible, the homes should be occupied by municipal employees, teachers, health workers, local working people, and longtime neighborhood renters ready for homeownership. Unlike the Island Cohousing project, we would strive here for full affordability—a range of subsidized homeownership opportunities and no market-rate houses. It would be a mixed-income neighborhood for all who were currently shut out of the Vineyard housing market, which included people making from 60 to 150 percent of median area income.

All homes would have permanent limited-equity deed restrictions to maintain their affordability in perpetuity. These restrictions provide for resale in such a way that, in the future, someone with the same relative income as the original purchaser will be able to afford to buy one of these homes. The use of these and similar mechanisms, like community land trusts that sell houses on property that they own and lease to the home-buyers, are key to the building of a pool of affordable housing over time.

Directly behind the property is an existing eight-unit affordable rental property owned by the regional housing authority. We determined that we could link the two, do improvements to upgrade the apartment complex, and make shared playgrounds and recreational areas. Rather than suffer a stark contrast between a nice new development and a slightly downtrodden old one, we would connect and integrate the two.

We created a ten-house cluster plan with parking on the perimeter and a pedestrian loop in the center. The plan included a mix of community and private space, including common storage for the ten households in structures at the parking areas, a solar-electric system for landscape lighting and common electrical needs, and common areas with a fire pit, a community garden, a playground, and a basketball court. The buildings were designed with good solar access (all roofs have large, unobstructed south-facing roofs) and high environmental and durability standards.

Before beginning the formal permitting process, we organized several public meetings for area residents. We presented the preliminary plan, answered questions, and listened to suggestions. The reactions were mostly positive, although the opportunity to be together caused some residents to point out some of the existing problems in the neighborhood.

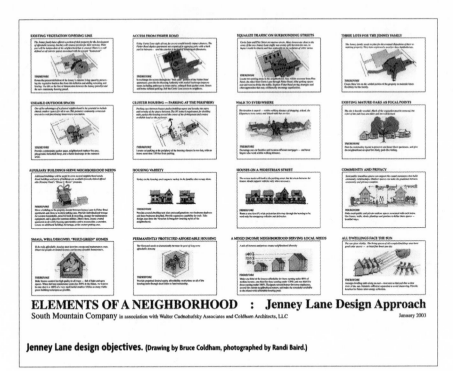

ELEMENTS OF A NEIGHBORHOOD : Jenney Lane Design Approach
South Mountain Company in association with Walter Cudnohufsky Associates and Coldham Architects, LLC   January 2003

Jenney Lane design objectives. (Drawing by Bruce Coldham, photographed by Randi Baird.)

Public officials were all supportive of the project. At an informal meeting with the Edgartown Planning Board, the only substantive remark we heard was, "Can't you squeeze in a few more houses?" No, we responded, this is what the property can bear.

To achieve the quality we hoped for and sell at the prices that would make the development affordable, we needed an average subsidy of $80,000 per house. I went to a wealthy philanthropic couple—former clients—who had been supportive of our efforts. They agreed to provide the entire package of necessary subsidies—$800,000 at one stop! Everything was falling together.

Wherever we went the project generated enthusiasm. The regulatory process involved application to the Edgartown Planning Board, which would hold a hearing and then turn the project over to the Martha's Vineyard Commission (MVC). Once we had received approval from the MVC, the project would be returned to the planning board for local approval.

At the first planning board hearing, it was clear that we had made a faulty judgment. We had expected the process to be a cakewalk, but this would be nothing of the sort. It turned out to be more like wading through a swamp with mosquitoes swarming our faces and alligators nipping our heels. The room was full of people bearing a litany of complaints. Some were rational and reasonable; others were not. Some we could accommodate; others we could not. There was acrimony and there were accusations. We had no idea where the hostility had come from. We were completely unprepared. We had not bothered to bring out supporters because we hadn't thought it would be necessary. The planning board was surprised, too.

The MVC hearings were similarly contentious, but now we were prepared. It was a strange dynamic. Three or four strong-willed opponents had aroused the entire neighborhood. I still don't know why. I think part of it was their sense that they had been overlooked and bypassed, as a neighborhood, by the town. We offered to help them with those issues, and did, but it had no effect. Their vociferousness seemed disproportionate to their expressed concerns, which were mostly about traffic and density. What else was happening? Were they worried about their property values? Did they disagree with the idea of affordable housing? They never said any of those things. They only said that the proposed development was too dense (although it was less dense than was permissible by right), that it would cause too much traffic (although we conclusively demonstrated that it would have little effect), and that it was out of character with the neighborhood (which was a matter of opinion, which we disagreed with, and for which, in any case, there are no standards). I was baffled by the inaccuracies of their communications and by the character assassinations that peppered their discourse. We brought in people to support the need for the project, as well as some people from the neighborhood who spoke in support, but mostly we had to just listen to the tirades, do what we could to make adjustments, and let the rest roll off our backs. After several lengthy hearings and a variety of delays, the MVC approved the project unanimously, with minor conditions.

We returned to the Edgartown Planning Board. I was worried about the Jenneys. This was taking much longer than any of us had figured, and some of the testimony was insulting to them. Would they grow weary and back out? No, their commitment remained firm. In the end the planning board, faced with a project they supported and an angry group of townspeople, did a superb job of devising conditions that in no way diminished the project (one actually enhanced it) but that defused and partially addressed some neighborhood concerns. They voted unanimously to support the project with the conditions they had imposed.

One silver lining of the controversy was that the residents formed a new neighborhood association. Town departments agreed to work with the association to solve some of the problems they had identified. Nevertheless, a few malcontents decided to appeal the planning board decision. When they filed their appeal, the *Vineyard Gazette* wrote an editorial that chastised the opponents and said, "The future of our Island community is most often decided in small steps—a board vote here, a commission approval there. Such decisions should be made after fair and open debate, and with plenty of listening. But appropriate progress is jeopardized if individuals refuse to recognize the legitimate outcome of a thoughtful process."[8]

Ultimately, I am sure that Jenney Lane will be a wonderful little neighborhood tucked within another.

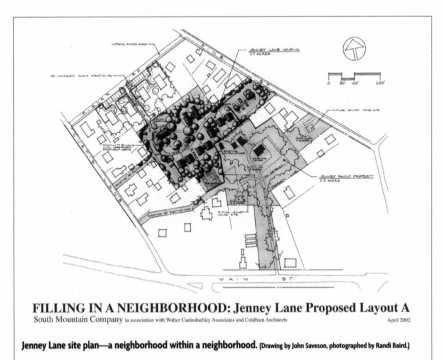

FILLING IN A NEIGHBORHOOD: Jenney Lane Proposed Layout A
South Mountain Company in association with Walter Cudnohufsky Associates and Coldham Architects   April 2002

Jenney Lane site plan—a neighborhood within a neighborhood. (Drawing by John Saveson, photographed by Randi Baird.)

Neighborhoods

In 1960, when I was in the sixth grade, I went to the Flood School in Menlo Park, California. Adjoining the school was a large subdivision, called Suburban Park. The subdivision had a billboard on Oak Grove Road, near the entrance, that read, "Suburban Park—New houses as low as $23,500—Low down payments—FHA loans available." The postwar housing boom was still in full swing. Suburban Park was a maze of winding roads and cul-de-sacs full of small ranch-style bungalows dominated by two-car garages. The houses were cheek by jowl, packed in about eight to the acre.

We lived two miles away, in a lush wooded area of half-acre and larger lots. Our contemporary house was set back from the road. The gracious backyard was fully surrounded by a six-foot wooden fence. It was a lovely, safe, verdant, quiet neighborhood. A few friends lived nearby, and we hung out at each other's houses. But our favorite place to play was at our friends' houses in Suburban Park. It was lively there. The school playground and playing fields were nearby. The streets were full of kids. Stuff was happening. It was a neighborhood.

Building neighborhoods is about more than just building houses and landscapes. Without vital neighborhoods, without community, without the kind of development that encourages people conservation, essential qualities of place cannot be preserved or restored. Someone once said, "We set out to change the world, but we had to settle for changing the neighborhood." That may be just enough.

South Mountain Company wind turbine. (Photo by Tim Mathiesen.) **Inset: Readying the turbine for the raising.**
(Photo by Ryan Bushey.)

⌈ 8 ⌉
PRACTICING COMMUNITY ENTREPRENEURISM

When we first came to the Vineyard it was to do a job—to build a house for my parents in the town of Chilmark. The people we met chuckled when we told them why we were there. "Move right into town, from outside, and build a house? No way. You gotta be Herbert Hancock to do that. He runs the town. He's the main builder. He's the only one who gets a permit."

Along with being the primary builder in town, Herb had already been a selectman for eleven years. He was also an assessor, a member of the local board of health, and the building and zoning inspector. Whatever went through town hall went through Herb. Alarmed by what we heard, we prepared our plans carefully before we went to see him. Contrary to the rumors, he was gracious and welcoming. He made short work of the permit and generously gave us useful advice about the mysteries of horizontal wind-driven rain and the complexity of island soils. Our experience with Herb was a lesson. It's not uncommon for those without authority to fear or suspect those who have it. But more often than not, municipal officials are simply people who are dedicated to their community. If we question rather than declare, understand the view from the other side of the table, and search out commonalities, collaboration is more likely than conflict.

More likely, I said.

Recently we built a house for conservation-minded clients who wanted their house to be a net energy producer; that is, it would produce more energy than it consumed. They had acquired a large property of significant ecological importance in a highly visible location, and they had treated the land and the development process with great sensitivity. Most of the land was put under permanent conservation restrictions, and our clients agreed to fund the Nature Conservancy to conduct a long-term sand-plains restoration project on the property. The building program was modest for people of such substantial wealth. The town and the neighbors were appreciative.

We proposed to erect a wind turbine to satisfy the energy needs of the house. We assumed that the goodwill that had been created, the large size of the property, and the fundamental logic of the proposal would carry the day. True to the lessons we had learned in the past, my colleagues and I studied hard, marshaled our arguments, and delivered them as persuasively as we could. But our application for a permit under the town's wind energy bylaw, which fully met all the criteria, failed to budge the planning board. They were unconvinced—application denied, unanimously.

Wind energy is part of our regional heritage. In the 1700s wind was used all over the region to pump seawater into evaporative ponds to produce salt. Wind energy powered local gristmills through the 1800s. Windmills all but disappeared for much of the mid–twentieth century, but after the energy crisis of the mid-1970s, interest in renewable energy fueled a resurgence of electricity-producing turbines. There were more than a dozen wind turbines on the Vineyard: an old recycled Jacobs machine from the post-Depression Midwest, the local utility's useless Department of Energy prototype, our own first-generation Enertech at Allen Farm, and a host of other jury-rigged wind catchers. In terms of effectiveness the results were mixed, but the turbines didn't seem to offend anyone. In fact, when we took down our machine at Allen Farm (after too much patching, fiddling, and replacing), the same people who had groused when we put it up protested more loudly—they had grown to love it! Fishermen at sea used it for bearings. Motorists driving by got an instant read on which way the wind was blowing. Imaginations were captured by the mesmerizing effect of the turning blades and the gratifying sense of what they were doing.

Most of those machines are long gone. In the intervening years, impressive innovations have led to a new generation of quiet, effective, gracefully designed workhorse turbines to replace the experiments of our forebears. So why was our application denied?

I wondered whether it had something to do with the well-publicized, hotly debated Cape Wind proposal to create a large offshore wind farm at nearby Horseshoe Shoals in Nantucket Sound.[1] The Cape Wind project is seen by some as the industrial development of a pristine area. To my thinking, the Sound is already industrialized by the pollution emitted from the Sandwich power plant, located at the end of Cape Cod Canal, and by the ships that bring the oil that fuels it. Just a few months ago, one of them spilled ninety-eight thousand gallons of oil into Buzzards Bay. The company that owns the ship was assessed one of the largest criminal fines ever for an oil spill. That kind of commercial activity is incompatible with other maritime uses and local ecology. A wind farm is not.

Cape Wind opponents predict a litany of problems, and they are strident about the issues. They call themselves environmentalists while vilifying a project that will significantly reduce air pollution and greenhouse gases (equivalent to taking 160,000 cars off the road). The real crux of the controversy is aesthetics. Many people believe Cape Wind will be a visual nightmare. Others, including lifelong sailor and local maritime historian Tom Hale, think "a wind turbine is as graceful a man-made structure as exists anywhere."[2] Tom has devoted his life to ships under sail. For him, perhaps a wind farm is like a harbor full of tall, slender masts, an expression of humankind's ability to harness that which is free and plentiful without doing damage to those things we treasure.

Some things can't be effectively argued. Aesthetics is one. The painting that hangs on my wall may well be appalling to you. It's especially hard to argue the aesthetics of something we can't even see, like Cape Wind. But we *can* look at the history of similar installations for indications of the effect the aesthetics may have.

When an eighty-turbine wind farm was proposed many years ago, just 2.4 miles off the beautiful sand beaches of Blaavandshuk, Denmark, a region that derives 90 percent of its income from tourism, there was widespread protest. Another proposed wind farm in the region brought nineteen hundred complaints. The protesters lost the battles. Both projects were built. Today 20 percent of Denmark's electricity comes from the wind.

One of the protesters was Jan Toftdal, the director of business and tourism in Blaavandshuk. Today Jan Toftdal is a big supporter. He sees no detrimental effects, and says they are getting new tourists who are attracted *because* of the wind farm. Now the region differentiates itself from other destination resorts by promoting the notion of clean energy in a beautiful place, the synergy between the turbines offshore and the beaches onshore. Similar turnarounds have happened over and over, worldwide, for many years. Cape Wind, and the negative reactions to it, are nothing new under the sun. Just the usual.

I witnessed another such turnaround closer to home. Some time ago I drove through the town of Hull, south of Boston, with my then eighteen-year-old daughter and a friend of hers. We were there to visit an immense wind machine owned by the local municipal utility. When we drove around Hull Hill, it came into view.

My daughter Sophie gasped, "It's huge. Scary."

We parked just steps from the machine and walked toward it. The tower is 165 feet high and the blades extend 75 feet above that. It is almost noiseless—it makes a gentle whooshing sound. This machine has sixty-five times the generating capacity of the one we had proposed on the Vineyard. As we walked away, we turned and stared back at it.

Sophie said, "It's quite beautiful, isn't it? Especially because of what it does."

Perceptions can change in a heartbeat.

The Hull machine is right on the beach, adjacent to the high school and a residential neighborhood, and in plain view of downtown Boston. It has been so successful that the town wants to build more. They polled the residents who live in the shadow of the turbine. Of the 500 respondents, 480 supported more turbines. That's 96 percent. When are 96 percent of people in favor of *anything*?

This degree of support is a common reaction, worldwide, in areas that are making a serious commitment to wind energy. As opponents battle to undo the Cape Wind proposal, many others are battling, as author Bill McKibben says, "to see them not as industrial eyesores, but as part of a new aesthetic. The wind made visible. The slow, steady turning that blows us into a future less hopeless than the future we're steaming toward now."[3]

It's tough to argue about aesthetics. But it's tough to argue with history, too. And even tougher to argue with the vision and the courage that it will take to free ourselves from oil pollution and oil wars. Cape Wind is one small but essential step in that direction.

We had tried to discuss the aesthetic and philosophical issues with the local planning board, but it had had no effect. As we spoke with them, it became clear that they believed the turbine we had proposed would be an aesthetic blight and would cause unforeseen problems. In response, our clients were willing to agree, although to no avail, to a periodic review that would result in the removal of the machine if there were unsolved difficulties.

Neighbors had mixed views. Some were vehemently opposed, others were cautiously intrigued, a few were supportive. Twenty years after an oil embargo and lines at the pump, in the midst of a war in Iraq, there was little mention of the symbolic and practical importance of the proposal, except from a few passionate advocates. The planning board made room for good dialogue and listened attentively, but stuck to their view. We told the members we disagreed with their decision, thanked them for conducting a thoughtful and considerate inquiry, which they did, and left the door open to return.

In the aftermath of the planning board's rejection, we wondered what we could have done differently. We resolved to do something to change the tide. A major obstacle had been the lack of a real-life local model that planning board members could experience. With that in mind, the owners of South Mountain decided to propose another such machine, in our own town and on our own property, in a less visible location. We decided to invest the time and money to make the model that, we believed, the community needs.

This time, the local and regional regulatory agencies were unanimously supportive. No neighbors objected, and many applauded. We recently raised the machine and have made the island aware of its presence. It's there to view. Perhaps this will pave the way to a positive decision for our clients when we return to propose, again, in a somewhat different location, the wind turbine that was previously rejected.

By installing the wind machine on our property, South Mountain sends a message of commitment to the community. When the people who work for a company own it, there is powerful incentive for the company to invest in improving the place where its owners live. Owning a business is like owning a home or a farm; it inspires a strong sense of responsibility and can, in my experience, also inspire a strong connection to the place where the business is located. When a community of people own the business, there is a collective self-interest that drives us to give back to the region that sustains us, to share our wealth and expertise to help build a stronger place.

This is community entrepreneurism—the commitment of business to bringing new ideas, investment, and problem solving to a local community. This has become the seventh cornerstone of our work.

Wearing Many Hats

Young people in the United States grow up with a political system that bargains back and forth about minor adjustments and never considers an actual overhaul. Swing a little left, swing a little right, one step forward, two steps back. In our huge, increasingly diverse nation, small changes seem to have little impact, and the big issues don't get tackled in significant ways. Participation in political life is low. In the absence of big changes in a big place, perhaps big changes in many small places can substitute. One avenue toward such change is the collaboration of three types of interdependent entities: (1) democratic, place-based small businesses; (2) local governmental agencies; and (3) charitable nonprofits that are dedicated to specific geographical regions. All three can bring different attitudes, modes of working, and funding sources. The intersection is dependent on strong community interest among the three and an involved, engaged citizenry.

Small businesses committed to a locale have a natural interest in creating a better business climate. Along with good commercial terrain, employees want a better place to live and raise their families. Businesses can act quickly and decisively, and they are able, if so motivated, to share the profits they earn.

Well-run local government can provide a broader platform through which to connect a particular business or undertaking to the citizenry. Government generally has a longer view and slower pace than business.

Private nonprofits, such as church and ecumenical associations, social service agencies, conservation organizations, housing groups, support groups, and a host of issue-oriented alliances and coalitions, can afford an even longer view, filling the gaps where government is unable to act. They facilitate volunteerism and effective citizenship. As people live longer and healthier lives, more and more people have the time and inclination to participate in helping to solve our social problems.

In small communities people often wear many hats and work on several sides of the table simultaneously. This can lead to conflicts of interest, long-standing feuds, and small-mindedness, but it also can lead to wonderful synergies. Small towns where people know one another in different contexts have built-in safeguards—family connections, business associations, and the ever-active rumor mill—that help maintain balance. When you can't hide, there's more incentive to behave. People learn whom they trust, whom they can work with. The wearing of many hats creates a more informed citizenry and greater possibilities for collaboration among business, government, and nonprofits. The three sectors can weave together their different perspectives and abilities to energize positive change.

In our company, one year our thirty employees included the chair of the regional planning commission, the vice-chair of the regional housing authority, two board members (including the chair) of the Island Affordable Housing Fund, one town conservation commission member, two members of town zoning boards of appeal, and many other civic and local government participants. These individuals bring the community into the company and the company into the community.

Greenough House

In 1994 a woman named Connie Sanborn called me. Her grandparents had moved to the Vineyard in the 1950s. After her grandmother died at the end of the decade, her grandfather bought a large old house, converted it into five apartments, and set up a nonprofit organization whose purpose was to maintain the house and make the apartments available at affordable rents to elderly women of limited means. He named it Greenough House, honoring his late wife by giving it her maiden name. He died in 1963. When Connie moved to the island in 1970, she and her husband took over management of Greenough House. There was little

Greenough House from the street Connie used to avoid. (Photo by Derrill Bazzy.)

money in the fund her grandfather had left with the house, but Connie was devoted to carrying on its legacy.

Now she faced staggering oil bills that threatened the viability of the place. She asked if I would come look at the house and determine whether anything could be done to reduce energy costs.

It was a beautiful old rambling Victorian. The apartments were warm and cheerful. The women who lived there seemed happy to be there, and friendly with one another. The house was grand but dilapidated, in desperate need of serious maintenance and restoration. Connie and her son couldn't keep up with it. She was demoralized and couldn't even stand to look at the house, remembering its former stature. She drove on another street when she was in the area in order to avoid being reminded of the house's sorry state. As the bills mounted, she was having trouble paying the mortgage, although she was fortunate that the bank holding the mortgage was sympathetic and flexible.

I concluded that doing energy conservation improvements would be like tossing a toothpick to someone fighting a bear. A major overhaul of both the building and the institution was needed. I suggested to the South Mountain board that we develop a plan to reenergize Connie's small and inactive board, raise money and community support, and put

Greenough House back on solid ground. The board supported the plan, and our support gave Connie new hope and energy.

Richard Leonard, CEO of the bank that held the mortgage, also championed the effort and agreed to serve on the newly expanded Greenough House board. He became one of the most important advocates for the project. Eventually the house was beautifully renovated into seven apartments and ownership was transferred to the regional housing authority, which maintains and operates it, and the place is a joy to behold. Connie no longer has to drive on the other street.

South Mountain did no design or construction on this project. All we did was get the ball rolling. One of our owners, Derrill, who was also a member of the regional housing authority, remained on the Greenough House board, stayed with the process through thick and thin, and was the prime mover, along with architect Kate Warner, in bringing the project to completion. South Mountain Company had only to prime the pump.

We learned that by stimulating cooperative partnerships among local businesses, nonprofits, and public agencies, we can make things happen that exceed what we could accomplish on our own.

Assessing Need

Our work on affordable housing differs from our custom work in this way: our custom clients come to us, asking us to fulfill their dreams, while our affordable housing work is mostly self-generated—we identify needs and create projects, proactively, as community entrepreneurs. Over time, our focus has expanded from buildings and technology to community planning efforts. In our small island community, we're finding that the need to create a new story is compelling, that the rewards for doing so are great, and that the potential to make progress with affordable housing, and preserving community, is within reach.

For several years now, I have chaired the nonprofit Island Affordable Housing Fund, and another of our owners is among its sixteen board members. The fund has established itself as an umbrella organization that raises money and distributes it to worthy projects and programs. Our company has invested time and money in this organization because we believe it creates solutions. The fund was responsible for the only in-depth regional housing needs assessment that has ever been conducted on the Vineyard.

When we began that inquiry, however, I worried that it might just become just another study sitting on a shelf gathering dust. I thought of the story of the sheepherder who was tending his flock in a field at the edge of a country road in rural Wyoming. A brand-spanking-new SUV came flying down the road and screeched to a halt, and the driver—nattily dressed—hopped out and approached the shepherd. He looked at the flock and said, "If I can guess the exact number of sheep here, will you give me a young lamb?"

The shepherd looked over the sprawling herd, stroked his chin, and said, "Sure, give it a whirl."

The young man connected his notebook and wireless modem, did some computer acrobatics, turned to the shepherd, and said, "Looks like 1,586 sheep here."

The herder said, "Whoa. You hit it right on the nose. Unbelievable. Take your pick."

As the man picked up a young lamb and turned to go, the shepherd said, "Wait a minute, son. If I can guess your profession, will you give me back my lamb?"

"Sure," the young man said.

"You're a consultant."

"Exactly. How did you know?"

"Pretty simple. You came here uninvited and charged me a fee to tell me something I already knew."

I thought the housing needs assessment might be just such an instance of paying good money to learn what we already knew. But the study—which was prompted by towns, donors, funders, and others asking questions about the nature of the housing problem and the profile of the need—told us far more than we knew. It portrayed a crisis larger than we had imagined and painted a picture that even skeptics could not deny.

My own skepticism about the needs assessment took a knock. Our weekly town paper, *The Broadside*, printed this: "Mr. Abrams tells us of a donation to the cause from a well-wisher who specified that his $25,000 be used to conduct an affordable housing needs assessment. Mr. Abrams says the search has been narrowed to two firms and a final choice is imminent. People in public life don't look back, but newspaper editors do. Last May, the record shows we asked John if it would help the cause of affordable housing to make a housing needs survey. His reply, 'When the river is overflowing its banks, you don't measure the water, you start filling sandbags.' Is it not wonderful how a well-directed check can change minds?"[4]

Well . . . what can I say? He's right. Community entrepreneurism entails taking risks, taking public stands, and learning as you go, often under the close scrutiny of local naysayers. Community entrepreneurs open themselves up to all of the vagaries of public discourse and debate. On occasion we have been taken to task on less accurate accounts. It's apparently true that no good deed goes unpunished, at least by a few. We are deeply involved in all aspects of the housing campaign—private, non-profit, and public. Some feel this is an elaborate "South Mountain Conspiracy," our own little corporate scandal. A conspiracy to do what, I wonder. Solve a problem? A few think we make money at it. If they can figure out how, I wish they'd let us know. There's nothing we'd like more than to make money providing good affordable housing, but we haven't managed it yet. The potshots just come with the territory. They are harmless at worst, and they keep us on our toes.

As it turned out, the needs assessment, which was called *Preserving Community*, has been particularly helpful with potential donors. As our influx of visitors and new residents brings community change, it also brings new wealth to harness. We need not be shy about developing tools that will enable us to mine these resources.

This community entrepreneur has been able to move forward on projects of interest only because of the supportive island-wide public/private partnership that has emerged. We all are working together to assemble a mosaic of inventive, scattered-site, small-scale, high-quality housing solutions that honor the beauty, the historical development patterns, and the rural past of the region. Unique Vineyard programs, such as House Moves and Rental Conversion, are emblematic.

House Moves

When land gets scarce and pricey, its value can far exceed the worth of the buildings located on it. People begin to tear down perfectly good houses to make room for new construction. With sufficient creativity, homes that owners no longer want can become an unusual resource.

Several years ago, the Island Affordable Housing Fund began to get calls from people wishing to donate their houses if we could relocate them. Regrettably, we had to decline, because it's far too expensive to develop land, move a building, renovate, and update. Then it occurred to us: the funding mechanism is right there, embodied in the houses! Now when we field such offers, we say, "Here's the deal. You donate your house to the nonprofit fund. We'll get your house appraised to determine the value of your charitable donation and we'll remove your house if you will contribute *in cash* the gross tax benefit you will receive so that we will have the money it takes to do the work."

The concept takes people aback at first, but after consultation with their tax professional, most agree, as long as they are in a position to take advantage of the tax breaks, which can be used over a period of five years. Recently one substantial property was assessed for $244,000 for the house only, not including the land below it. The total tax benefit to the owner who donated the house to us turned out to be 52 percent, and we received a check for $127,000. Another was a beautiful 1841 Victorian house that was in splendid condition but was too small for today's extended-family vacation life. The cash donation that accompanied the donated building came to $80,000. The owners *pay only what they'll save*, and in addition, they also save the cost of demolishing and disposing of these buildings, which is not inconsequential, and they provide a service to the community. Everyone comes out ahead.

The Town of Edgartown, the nonprofit Island Affordable Housing Fund, and South Mountain recently collaborated on a pilot project. Four homes were moved to four adjoining parcels owned by the town. The homes are equipped with perpetual limited-equity deed restrictions to guarantee that they'll be affordable forever, and they were lotteried off at remarkably low prices to carefully qualified local residents. One of the recipients was a dignified, white-haired gentleman in his fifties who had

An historic Victorian moving down the road. (Photo by Derrill Bazzy.)

worked in the school system for many years. He had lost his rental when his landlord had decided to sell. Purchasing a house had seemed far beyond his means, and he thought he would have to leave the island, where he had lived for nineteen years. It was a great moment when his name was drawn out of the hat, thereby selecting him to receive one of these houses. He was overwhelmed with joy and relief and broke down when we spoke with him immediately after his selection.

Not long after I called him to ask him to testify in support of another project at a public hearing. He happily agreed. I told him I was planning to show the video that included our previous interview with him and asked whether he would find this embarrassing. He said it would be, but he'd be there anyway. He said, "You know, these days it's okay for a grown man to cry. I cried at the award, I'll cry again at the closing, and I'll cry the first time I walk through the door of my new home."

More and more houses are coming up for removal. A major problem is that it takes a long time to assemble a project, find land, develop it, and move the house. Owners aren't always willing to wait so long for removal. We hope to convince the towns to set aside land for storage of houses. Eventually we could have "used house lots," and when a young couple finally gets their land together and is ready to build, they'll be able to inspect the houses, point, and say, "We'll take that one."

The physical process of moving houses down narrow country roads and through tightly packed villages is a remarkably complex undertaking. Power lines, road width, overhanging limbs, the size of the house, the shape of the house, other traffic on the road—all have to be juggled. Sometimes you have to break a house into parts, move it, and then put it back together. Sometimes you have to cut it apart horizontally because it's too tall to fit under obstacles between its old and new locations; other times you have to cut it apart vertically, because it's too wide to traverse the narrow roads. Sometimes you have to do both. Only when you're lucky do you get to move a house whole.

At one point during a recent move, our job foreman, Peter D'Angelo, stopped by my office to talk about the dangers and difficulties of the job. We talked a while and then he left for the job site, where the house, which was already mounted on two trailers that had been moved out into a field, was to be split in half and rolled to its new location several miles away.

Late that morning I got a call from him.

"You remember what I was saying about the dangers?" he said.

"Uh-oh," I said quickly. "What happened?"

"Everyone's safe. But not the house. When we separated the two halves and moved the first half out, the second part toppled right off the trailer and crashed on its side."

After he'd reiterated that all were okay, I asked, "So what are you doing now?"

House topple: hard to believe the moment was captured on video! (Photos by Mark Lipman.)

"I'm trying to decide whether it's salvageable—whether to bother moving the sound half, whether to set a match to the whole damn thing, or what."

"Want me to come out?" I asked.

He hesitated, then slowly said, "Well . . . no, I guess not. We'll figure it out."

"You sound a little stunned," I replied. "I'm leaving now, okay?"

"Okay," he said. We hung up, and I headed for the door.

The scene was surreal. Because the move required power, telephone, and cable television lines to be disconnected and lowered, the site was jammed with utility trucks, linemen, police escorts, and workers. They were standing around in small groups staring at this strange, two-story half-a-house lying on its side. Peter and Mike, the mover, had already figured out a scheme to right the house, repair the damage, and move it.

There was little talk about what had actually happened, and no talk about who was to blame. Everyone began to put things in order for the move, like cowboys pulling the herd back together in the quiet moments after the end of a stampede. The utility guys headed for their trucks, the cops moved out to the road and switched on their flashing lights, and the tractor started up and the other half began to roll.

Not long after I left, Peter told me, the local newspaper showed up. When pressed by the reporter to make a tragedy out of a mishap, as reporters are wont to do, Peter quipped, "The closest thing to a real tragedy this morning was that while we were otherwise engaged, the neighbor's dog stole our supply of fresh-baked Humphrey's donuts."

As I said, community entrepreneurs are always learning, and with learning comes a certain vulnerability and exposure. Our mistakes are right out there in public view. Which adds a certain sweetness to our occasional successes.

Rental Conversion

The idea of using existing housing stock to its full potential led to another entrepreneurial strategy: stimulating the conversion of under-utilized seasonal rental housing into affordable year-round rentals.

In a resort community like ours, many investment properties rent at high prices for the three summer months, in that short time generating a full year's worth of income. The regional housing authority let the community know that if property owners would commit their house to year-round affordable rental, they would subsidize the difference between the income so generated and the income the owners had been getting from summer rentals, and they would manage tenant selection, too, assuring that the rental would be filled.

Willing rental owners came out of the woodwork. Within a few years fifty families were in stable year-round rentals. Many of them were people who had been moving twice a year for many years, in what is known locally as the "summer shuffle." Winter rentals are easy to come by, but come summertime many folks are forced by prohibitively expensive seasonal rents to move in with friends, camp in the woods, or even leave the island temporarily. One of the conversion program's tenants recently said:

> My son is going to be five in two weeks, and he's lived in twelve different houses in his lifetime, twelve different places. Moving in here has helped us bring cohesion to our lives. You know, you have to have the same place to come home to every night. And somebody this size has to know where that home is going to be, and that every night it's going to be the same place, I think. I mean, they have that right.[5]

Supporting this program's rental subsidies requires substantial private fund-raising. To date, this program has been funded by an annual Island Affordable Housing Fund event called Houses on the Move, in which architects, builders, craftspeople, and artists make small houses—from jewelry to ceramics to doghouses to saunas to garden sheds to writers' shacks— that are all trucked to our Grange Hall for display and auctioned off.

In 2002, the first year, Houses on the Move raised $180,000. South Mountain's employee-owners designed and built a seven-foot by nine-foot writer's shack out of found materials, including roofing made of the discarded aluminum printing plates from the *Vineyard Gazette*, which you could read inside through the skip sheathing of the roof. It was fully furnished and equipped with a well-oiled Royal manual typewriter. It sold for $40,000.

A ceramist fashioned a wonderful little pickup truck, overflowing with one family's worldly goods, including the dog. Symbolizing the summer shuffle, it sold for $3,000.

Successful community entrepreneurism gives towns and regulatory agencies the confidence to take larger steps. It gives public-spirited developers and builders the sense that if they are creative and willing to step

The Writer's Shack. (Photos by Brian Vanden Brink.)

forward, their projects may find acceptance. Vineyard towns are now actively seeking to develop town-owned lands. Zoning incentives are strengthening. The regional housing authority has been energized. A community housing trust has formed to accept, hold, and manage land for affordable housing in perpetuity. Private donations are coming into the Island Affordable Housing Fund's "Raising the Roof" campaign. The business community has formed the Business Initiative for Housing Solutions and citizens have united behind Islanders Helping Islanders. Local churches have formed the Housing Ecumenical Action Team, which has incorporated an active nonprofit affordable housing development entity. Habitat for Humanity has a busy local chapter. Schools are developing teacher housing. It is extremely gratifying to play a part in this gathering momentum.

Commerce and community are united in this effort. Pessimism and lack of will have been, in large measure, overcome. This is not to say that there is no longer any not-in-my-backyard sentiment. Sometimes such expression is vigorous. Mostly, though, there is a solid new consensus building around a core of community support. Progress on these efforts cannot come fast enough. The community loses people every year due to the lack of housing solutions. As important as each individual project is the larger unfolding of a new direction for the community.

In this process of long-term reorientation and the nurturing of an emerging consensus, community entrepreneurism has become a vital, restorative force.

Everything was made for the room: lighting, table, and benches. (Photo by Brian Vanden Brink.)

9

THINKING LIKE CATHEDRAL BUILDERS

During the last days of January 2000, South Mountain Company's twenty-five employees, along with several friends and planning experts, spent two days hunkered down together, thinking about the future of the Vineyard. Our goals were to sketch a future we would like to see and to decide what commitments we, as a company, were willing to make to help achieve it.

The prelude to this exercise was our sense that Martha's Vineyard is being overwhelmed by its own desirability and prosperity. Land values are rising and community life is changing. Most people are acutely aware of the changes, but our political leadership has not formulated an effective collective vision of the future. Instead, our posture remains defensive. We act as if we are under siege, weakly trying to fend off the inevitable.

We asked ourselves, "But is it inevitable? Can we imagine a future that would make it possible for our children and grandchildren to enjoy the Vineyard the way we have? Can we muster the strength and resolve to tackle the issues and forge a satisfying tomorrow that serves the interests of all? Perhaps . . ."

We invited a few people from outside the company to Future Sketch, as we called the meeting, to broaden our perspectives. We gathered early on a Friday morning. After introductions, author and historian David McCullough, one of our guests, opened the day. He spoke about the Chagres River, which was the major obstacle to the building of the Panama Canal, but which was eventually used in a simple but ingenious way to become a part of the overall engineering solution. He related this to the "river of money" pouring into the Vineyard, undoing a way of life. He expressed two ideas that became central to our discussions:

1. "We must redirect the river of money [that causes such harm] to restoration of community."
2. "Our future is a design issue—it should be the result of intent rather than circumstance."

With these two themes echoing, we started by sharing memories of the past. After contrasting the characteristics of the place we love with the nature of the pressures that were changing it, we began to consider the components of a positive future—one that might harness the forces of change to allow the Vineyard to continue to be a place we (and the generations to come) will love.

The Results

We assembled a collection of ideas that might guide us, and inspire others, to chart a course that would make a difference. Some of the ideas that emerged were more developed than others. Some were tied to initiatives that were already in place and in need only of encouragement or better funding. By the end of the session, we had identified eight crucial areas:

- achieving political unity
- creating the Vineyard Institute
- promoting adequate and appropriate housing
- preserving and enhancing rural character
- supporting the new traditional economy
- making a transportation system that works
- committing to environmental stewardship
- maintaining cultural traditions

In each of the eight areas, we characterized the problem, suggested solution concepts, and identified actions to be undertaken. Having worked through the eight issue areas and come to broad consensus con-

The Future Sketch gathering. (Photo by Derrill Bazzy.)

clusions, we shifted our focus. We asked ourselves, "Given the vision of the Vineyard that we see before us, what can we, as a company, do to further and support the future we've identified?" The discussion was lively. Ideas flowed easily. The result was a series of fourteen actions to pursue as a company, including developing internal regulatory processes to assure appropriate building scale and good land use in our work; devoting a larger portion of our profits and pro bono time to community efforts; making a greater commitment to affordable housing projects; intensifying our efforts to use renewable energy and materials and to eliminate waste; aligning our mission with the Future Sketch findings; and sharing our conclusions with others.

Through this discussion, there was a discernible, cohesive force linking us, a powerful community of interest. We agreed that we had not crossed the line of "fouling our own nest," that we could still steer in the direction of a longer view and encourage effective solutions over time. We understood that our problems were the same as those facing all beautiful and desirable places. We recognized that already we had some elegant successes to point to, and we concluded that it's as important to remember what's been accomplished as to identify what needs to be done. We decided that it would be essential to create good models that would make radical new approaches seem commonplace. The Future Sketch undertaking itself was one such example.

Buckminster Fuller once said, "You never change things by fighting the existing reality. To change something, build a new model that makes the existing model obsolete."

It's unusual for a small company full of carpenters, woodworkers, designers, and office people to take two days to consider the future together. There was, in fact, some skepticism and grumbling beforehand. The results, however, were satisfying (the full Future Sketch report is reprinted as appendix 3).

The Dumptique

It was on the first Earth Day, in 1970, that Vineyard forester, conservationist, ox driver, tinkerer, and dump picker Bob Woodruff took a group of high schoolers, hauled his oxcart along Beach Road, and filled it up five times with trash, bottles, and rubble. They decided to separate that which could be used from that which couldn't when they took it to the dump, and recycling was formally born on Martha's Vineyard. Soon a concrete bin was erected at the West Tisbury dump (this was in the prelandfill days when dumps were dumps) for glass recycling. Bob called it the "aggression relief bin" because everyone would stand and wing their bottles in one by one. In the late '70s a crude building was erected to shelter recycled newspapers. People started leaving boxes of clothing, someone put some hooks on the wall and began to display the clothes, and townspeople began to stop by and pick up the good stuff.

A woman named Dot West became the voluntary supervisor of the place in the early 1990s. She organized and ran the operation, and the recycling shed became Dot's Boutique, a true West Tisbury institution and destination. Dot became quite possessive, began to padlock the door, got into a dispute with the West Tisbury selectmen, and stormed off in a huff. Sometime later Tim Maley decided it ought to reopen, and with Jean Wexler and Ginny Jones he cleaned it out and opened the doors of

The Dumptique. (Photo by Peter Rodegast.)

the newly christened Dumptique. Jean has run it ever since, and for years it has been the first stop for fine clothing, used crutches, immaculate curtains, cast-iron cookware, and all manner of good stuff discarded from one person's life and useful in another's. My co-owner Billy says most of his best flannel shirts come from the Dumptique, my neighbor Craig wears a beautiful blue sweater he selected on one of his missions, and many of the island's Brazilian residents hunt throught the Dumptique's wares for presents to send home. In recent years it continued to thrive, but the gnarly little building fell increasingly into disrepair.

In January 2000, at the Future Sketch, the sixth of South Mountain's fourteen resolutions was:

> *Offer to improve the West Tisbury Dumptique and add a building materials exchange.* Take this on as a South Mountain project. Discuss with the Refuse District, the West Tisbury Board of Health, and the Dumptiquers to determine needs. Offer to assemble a plan and proposal for the Dumptique and a building materials exchange. If it is embraced (or even encouraged), do it!

Four years later Peter Rodegast, one of my co-owners, brought this up at a board meeting. There's talk in town, he related, of putting together an effort to restore the Dumptique to its former grandeur, but nobody's actually stepping forward to make it happen. Should we help out?

Yes, everyone responded, let's take it on. We agreed that Peter should lead the effort, that we would pay him to do so, and that we would contribute much of the material needed to do the job. Peter put together a plan with the Dumptiquers and organized townspeople to do a barn raising. For four days in September people pounded away at it, installing a new roof to protect the goods, skylights to illuminate the gloomy interior, and a fine front porch to give shelter from storms. Jean Wexler and her helpers were beside themselves with joy.

One more of our Future Sketch resolutions had come to pass. Next step: a used building materials exchange for all the superb building materials that get tossed out regularly.

Mostly we don't think about the issues we uncovered during the Future Sketch on a daily basis; we tend to go to work and do our jobs. The underlying principles are always there, however, guiding us quietly. We know they're around, like a rabbit scuttling in the underbrush. When I read through the resolutions we reached, I realize that many of the ideas and suggestions have been implemented, to a lesser or greater degree, in the five years since. The Future Sketch was an important passage for the company. Perhaps its most immediate impact was that it facilitated our thinking in generational terms about our work and our legacy.

Legacy Building

Our developing commitment to place and our policy of limiting our work to the Vineyard led to our efforts to think about the future of our region. It was a natural progression to begin to consider, in new ways, the future of our company.

How can we enhance the durability and the longevity of our buildings? How can our buildings contribute to the enduring qualities of this place? Can our company become an enduring participant in the future of this community?

Thinking like cathedral builders, whose work would not be completed in their lifetimes, has become our eighth cornerstone principle.

Some ancient cathedrals took centuries to build. Their builders had endless social, political, economic, technical, and climatological factors to contend with. The work continued as the cast of characters changed, the master builder died, wars were fought, and regimes changed. These remarkably complex undertakings somehow endured. The perceived rewards must have been great.

Whatever the personal rewards that accrued to each participant, cathedrals stand as important community anchors. The great ones evoke a sense of permanence and architectural grandeur, reminding us of a vision so strong it could be maintained over time.

In the case of a company, it's fair to ask why longevity and survival matter. Why not let the company die when it seems that it no longer has a purpose? To arrive at an answer we must ask other questions: What *is* our company? What is its purpose?

More than anything else, we see South Mountain as a community, and its purpose as a community will never be done. Its legal structure takes the form of an employee-owned cooperative corporation. But its function, its essence, and its purpose revolve around the web of relationships in its internal community and in the larger community of which it is a part, which supports it, and which it serves. The people who are part of South Mountain and the people whom it affects have ongoing relationships with the company. These relationships overlap, and I don't imagine, for instance, that there will ever be a particular day when everyone in the company will be ready to retire. There will never be a moment when everyone's houses cease to need maintenance, alteration, or addition. We will never complete the learning that can allow us to make better buildings and build a better economy. It is highly unlikely that we will look around someday and say, "Hey, we did it. All is well. We committed to each other and to this island, and now we're all fine and it's fine too. The job is done." Because these things won't happen, we are organized around the idea of maintaining and perpetuating our community for one another and for future generations.

Arie de Geus, who spent much of his career at Royal Dutch/Shell and is considered to be the person who originated the concept of the "learning organization," says in *The Living Company* that a company is *primarily* a community and that its purposes are longevity and developing its potential. To produce both profitability (a means to those ends) and longevity, he says, we must attend to the processes that build community. He goes on to say:

> Founders and managers of long-lived companies, a hundred years or more in the past, did not link their values to a particular product, service, or line of work. They knew, or sensed, that the life mission of a work community was *not* to produce a particular product or service, but to *survive*: to perpetuate itself as a work community.[1]

One of the key elements of community creation at South Mountain is our policy of hiring "future owners" as opposed to employees. We envision people who enter the company staying in the company and becoming owners of the company. We don't know what they, as the perpetuators, will do or what they will produce, but the essence of our collective enterprise will survive in them. Royal Dutch/Shell's ability to consider these things will probably serve it well in the years to come, when the age of oil is over. It is likely to evolve to the sale of other products and services. De Geus refocused the company on being a particular kind of company rather than doing a particular thing, thereby equipping it for the long haul.

Regions, towns, neighborhoods, and individuals are heavily impacted when a business closes its doors. Only forty years passed between the time my grandfather started his business and the time it began to be sold off and shuttled around. The community he had developed was left unprotected. Once we have succeeded at the job of creating community, our new job becomes its maintenance, enhancement, and perpetuation. When we build a business, we are building a legacy.

Peter Senge, author of *The Fifth Discipline*, the book that popularized the idea of the learning organization, writes about the difference between a company that is seen as a "machine to make money" and a company that is perceived as a "living being" with a heart and a mind. He states:

Seeing a company as a machine implies that it will run down unless it is re-built by management. Seeing a company as a living being means that it is capable of regenerating itself, of continuity as an identifiable entity beyond its present members.[2]

When the employees own the company, it becomes that kind of living entity from early on, and it is, we hope, better prepared for the journey to come.

The Sabbatical

I thrive on the clamor of work. It feels good when people poke their heads in to ask me a question while the phone is ringing, a dozen e-mails need replies, and the to-do list is long. But I like it when the dust settles, too, like before dawn on the day I'm leaving to travel somewhere. I've come over to the office from my home next door to collect my computer and a few last-minute items. My desk is clean and spare (for a change), most everything's done that needs to be, and I have extra time on my hands before I go. I've fed the stray cat and made the coffee, and now I can wonder what I forgot to pack. And sit for a few minutes, thinking, feet up on the desk.

During one of those early-morning musings years ago, I had a thought: when our youngest child, Sophie, goes off to college, maybe Chris and I could take a year away, and I would take a year's sabbatical from work. I didn't know what we would do or whether I really meant it, but the idea stuck, and it kept returning. As the time neared, it began to come into focus. It was clear to me that I wanted to write a book about business, but I had no idea what shape it would take. It also occurred to me that after many years of employee ownership, perhaps there was potential for the company to grow out from under me, to grow into more of an entity unto itself. My absence might provide the space and opportunity.

It would be the first time in twenty-eight years I'd been away for more than a few weeks. Planning for the sabbatical within the company began to make us wonder what would happen to South Mountain when my working days were done. It was to be the beginning of a long march toward succession, an expression of intent by the small workplace community to endure beyond my tenure.

Being away from my work has allowed me to write this book, and being away has allowed the company to do remarkable new work. It turned out that instead of a year away, we took two consecutive winters. I arrived in Vermont the first winter with milk crates full of books, a case of Jack Daniels with special "John Abrams Sabbatical" labels that was a going-away gift from my coworkers, and big questions about myself: Having been engaged in business so long, could I disengage, and sit, and focus, and write? If I could, would I like it? Could I leave the business alone enough that the kind of growth we were imagining would have room to happen? Would things go well without me, or would there be dissension and resentment? Should I just drink all the whiskey and forget about the rest?

I began to read, and it turned out that I kept reading—solid—for nearly two months, gulping down pages like a thirsty man who had just found water. Gradually I buckled down to work and found myself running down a series of blind alleys, but each time, before backing out, I scavenged and collected useful material. Occasionally I found alleys that led through. I was able to focus, I was having fun (some days), and I wasn't thinking obsessively about the company.

My primary diversion, as I wrote, was skiing for an hour or two most days. The joy of skiing is the search—for fresh snow, for new lines, for feelings of grace. Finding a line is the act of planning a descent while in the act of descent. Last year, on a snowy spring day in Stowe, Vermont, with my son Pinto, my grandson Kalib, and Kalib's friend Nate, I ducked off the Cliff Trail into tight trees. We caught a little north-facing pine-treed pocket where the sun had been refused admission and the snow was still deep and soft. We hadn't known this glade was there. We were lucky and hit it right for a sweet run. There's something sublime about picking your way through thickly wooded glades or crossing a high traverse on a powder day, looking for that perfect line. The likelihood of that perfect line is slim because the variables are huge—the terrain, the snow quality, the temperature, the wind, how you're feeling, how you hit it—but when you get lucky and find the line, when each turn is smooth

Sabbatical whiskey. (Photo by Randi Baird.)

and you're reading the terrain and looking three turns ahead, never seeing the tips of your skis but only the path you're about to create, your mind picking the routes and your body making the adjustments, the pleasure of effortless flow mixes with the thrill of exploration. Each line is new and different and unique; it is being invented as you go.

Writing this book has been like that for me. Early on, I was lost in a blizzard. As I began to understand what I meant it to be and where I was headed, and I felt myself arriving somewhere, it started to come into alignment. I gradually got my legs beneath me and learned to

maneuver—tentative turns at first, but with gathering speed and flu-
idity as I began to know the terrain and sense the outcomes.

It was good to be away.

Things were going well at South Mountain, but there was confusion
and stress. Some employees felt that they were overburdened and that
others weren't pulling their weight. Things fell through the cracks. The
management system we had devised to cover my absence was flawed.
Those troubles did not obscure the fact that when I returned in April, I
came back to a better company than the one I'd left. There was tremendous pride and good feeling. People felt that they had truly stepped up,
produced, and paved the way to new transitions. Indeed they had, and
they were ready for more. The time between the two sabbaticals was
devoted to picking up the pieces, gathering ourselves, making adjustments, and putting our heads together to prepare for round two. As we
approached phase two, the energy was palpable and the path was clear.

While I was off living a dream, I also had a chance to consider the company's future. Back on the Vineyard, an extraordinary group of dedicated
and gifted people was building on the cornerstones, making the foundation for that future. The development and growth the first winter had
been halting, slow, step by step. The second time through it was dramatically different, bold and decisive. We had replaced our eight-person management committee with a three-person executive committee consisting
of the people who had taken on the most responsibility. This group nav-
igated effectively.

My absence opened a floodgate, setting free a powerful stream of new
management approaches. A greater sense of responsibility and a deeper
understanding of collaboration spread throughout the company. A new
company was under construction. I was excited to return and lend a hand.

It took all of eighteen years for co-owners to assume fully the burden of
responsibility that my absence demanded. During those years, we developed a clearer sense of company values and a stronger set of relationships.
We may spend another eighteen years, or far less—who can say?—making
the actual transition from me, as leader, to whatever or whomever is next.
That's okay. Our job is to ensure that the values that are now embedded
are accompanied by the skill and the motivation to practice them over a

long period of time, without my leadership, and to go beyond where I have led. My time away helped us identify what needed work so that we could tackle those areas gradually and thoroughly.

Large organizations—corporations, governments, universities, professional sports teams—are expected to handle changing regimes and leadership. Small businesses are not; they are notoriously dependent on their founder. The first time I remember anyone at South Mountain saying, "What happens after John?" was at a board meeting in 1993, when I was still in my early forties. I'm certain it was not coincidental that it was also the first time we had consciously tackled the issue of defining our core values, our mission, and our purpose. That may have been the moment when we turned into what De Geus calls "a living company," although the process certainly had its roots back in 1987, when we restructured and became employee-owned. Although we didn't know it then, we were beginning to plan for succession and beginning to build a legacy.

If we begin to reduce reliance on the founder and leader of a company slowly and steadily, over time, we may be able to manage organic transitions instead of having transitions play themselves out in sudden and unexpected ways. People who are new to the process have time to watch and learn. As my co-owner Mike Drezner said:

> A group of people of goodwill and character can learn new roles and make worthy decisions. They can become facilitators and leaders. Making them full participants is a part of the journey, the transition to a different way of thinking. During the sabbatical period there was intensive structural and psychological change. New committees were formed. New attitudes developed.

The sabbatical, and the planning and internal growth it stimulated, was not the beginning of a subtle exit strategy but rather an unvarnished attempt to begin the preservation and perpetuation of a valuable community and to take it beyond its founder. It would be false modesty to downplay my influence and importance to the company, but it would be equally shortsighted to overlook the role I play as an impediment to the assumption of responsibility by and the professional development of others.

A Legacy Gathering

In the fall of 2003, Marjorie Kelly, publisher of *Business Ethics* magazine and author of *The Divine Right of Capital*, a hard-hitting analysis of our current brand of corporatism, gathered thirty people to spend an evening and a day discussing the issues of legacy and to launch the Legacy Project, which was intended to be an ongoing exploration of this subject. The fundamental question was, How can the values embedded in a socially responsible business be maintained when it is sold or when the founder retires?

Marjorie described the backdrop in a 2003 article in *Business Ethics*:

> It was April 11, 2000[,] when the legacy problem burst into view. That was the day the Ben and Jerry's board was forced by law to sell the premier socially oriented firm in America to multinational Unilever, against the wishes of CEO Ben Cohen. In the three years since 4-11, Bern and Jerry's has seen its social mission begin to seep away—Unilever has laid off one in five B&J employees, stopped donating 7.5 percent of profits to the Ben and Jerry Foundation, and hired a CEO Cohen did not approve of. It's been a wakeup call in socially responsible business circles, where preventing mission loss when a company changes hands has become the problem of the hour.[3]

She observed that we were entering a new era of socially responsible business, and that the founders' era is passing:

> Entrepreneurs have met the challenge of how to manage in socially responsible ways, but few even recognize the new challenge ahead: how to create the architectural forms that can hold social mission for generation after generation to come.[4]

The questions posed at the gathering were compelling to me because they were so closely related to the questions we were grappling with at South Mountain. Mostly, the discussion was about the kind of corporate entities we must create to encourage long-term social responsibility. Everyone there was familiar with the recent tendency of socially responsible businesses to lose their independence, and all were concerned that the important gains of the corporate social responsibility movement would be swallowed up and lost.

Leslie Christian of Progressive Investment Management in Seattle described her work on a new corporate structure. She wanted to create a Berkshire-Hathaway-type holding company that could be a guiding light for a new ethic. It would begin as a private company; then it would go public, chartered with the purpose of serving the public good rather than maximizing return to shareholders. The ownership structure would differentiate active from passive owners. Different share classes would have different voting rights. People who worked there and direct investors would get one kind of share. If you left or traded your shares, they became a different class of shares. According to Christian:

> A basic assumption about business is you can't responsibly invest without majority control. What would it be like to have a deep and trusting partnership without control? Control is a really old model. What are our deeper assumptions about how business has to operate?[5]

There were other inventive schemes proposed. There were also discussions about employee ownership transitions, and how difficult it is for people to make the cultural shift from being employees to being owners. I've watched these transitions over nearly two decades, and I've listened to reports of others. I've come to the conclusion that everyone is a leader, just as everyone is a designer, if they are brought to a seat at the table. The problem I see is the abrupt way in which the change usually takes place: the owner sells to the employees, and suddenly they're the owners. If the legacy begins as part of the early organizational planning, it can happen gradually, so that it's a continuum rather than an event. This allows time for training, for the development of decision-making and leadership skills, for the formation of social bonds, and for the emergence of collaborative management systems.

Someone at the Legacy Project meeting said that the reason most
entrepreneurs form companies is not to make money. They form companies because they need a job, and this is the kind of work they like to do.
John Logue of the Ohio Employee Ownership Center said, "The primary
legacy of employee ownership is the legacy of a retiring owner concerned about the economic security of people he worked with, the
people who built the business." So, let's put those two together. The
entrepreneur starts a business and begins to build it. That's what entrepreneurs do. He or she brings in people to help, with the idea that they
will become owners. When it's clear the business is going to be a success,
arrangements are made for the gradual sale of the company to the
employees. This is the path we have taken at South Mountain by chance
rather than by design. We threw in the ingredients and stirred the soup.
Only now are we coming to understand the recipe.

The Long View

Most small businesses that endure are family businesses that are passed
from parent to child. As family structure changes and our society offers
broader opportunities, however, it seems that fewer businesses are passed
down from generation to generation. In craft-based businesses like ours,
the employees *are* the business. The development of the business is synonymous with the evolution of the employees—our own unfolding,
blooming, and ripening at work. The business is a community, and one
of the essential reasons for its existence is to maintain the community it
has created. Employee ownership holds within it the seeds of continuity.
But we must plan for succession.

As I peer out at the American business landscape from my perch at this
small business, there seems to be all too little intergenerational thinking.
We have become a short-term culture, characterized by disconnection
and fragmentation.

Danny Hillis, the inventor of massive parallel computing, recently
designed an immense clock that will tick once a year, bong once a century, and chime once a millennium. The clock is now being built in the
California mountains. Its purpose is to illustrate a different way of

thinking about time. It is intended to work for ten thousand years, the span of human civilization to date. In Stewart Brand's book about the project, *The Clock of the Long Now*, the author says:

> Civilization is revving itself into a pathologically short attention span. The trend might be coming from the acceleration of technology, the short-horizon perspective of market-driven economics, the next election perspective of democracies, or the distractions of personal multitasking. All are on the increase. Some sort of balancing corrective to the short-sightedness is needed—some mechanism or myth that encourages the long view and the taking of long-term responsibility, where "the long term" is measured at least in centuries.[6]

This is how we're thinking about our community, our business, and our buildings. This is how we're beginning to think about problem solving, too. Big problems don't often respond to short-term bursts of energy. Problems take on a different cast when we put them into a long-term perspective. Hillis points out that difficult problems become impossible if you think about them in two- or five-year terms, as we usually do, but they become easier if you think in fifty-year terms:

> This category of problems includes nearly all the great ones of our time: The growing disparities between haves and have-nots, widespread hunger, dwindling freshwater resources, ethnic conflict, global organized crime, loss of biodiversity, and so on. Such problems were slow to arrive, and they can only be solved at their own pace.[7]

The long-term approach breeds optimism and resolve. From the short-term vantage point, sometimes it looks like we've crossed a line—there's no way we could ever go back and no way we could take steps big enough to matter. Sprawl is a good example. How could we possibly reverse or change the haphazard cycles of urban and suburban growth of the past several decades?

If we think in sufficiently long time frames, we can see how cycles of

development and restoration work. Take the Martha's Vineyard town of
Chilmark, for example. A little more than a century ago the town had
three thousand people and eight brick factories, and there was not a tree
in sight from one end of town to the other. The voracious appetites of
grazing sheep and burning brick kilns had turned forest to pasture. Today
the population has declined by two-thirds, there is no industry, and the
town is 90 percent wooded. It's an *entirely* different landscape—not better
or worse necessarily, but certainly far more diverse. Chilmark is a living
example of the extent to which landscapes and communities can change
over time.

That change happened by chance. Others happen by choice. Making
new landscapes and shaping new communities takes twenty-five, fifty,
even a hundred years. Here on the Vineyard we can begin the process of
restoration at the same time that we create a New Vineyard that can comfortably accommodate future growth—a New Vineyard made up of villages, farms, and wilderness like the old one, but incorporating what's
best about technology and the modern economy. We can't do this by
turning back the clock; instead, we must turn it forward. We need to
imagine a way that is economically robust—realtors need to continue to
sell, builders need to continue to build, property rights must be respected,
and everyday rhythms of life must be preserved. Can we imagine a future
like that—a future that does not turn its back on technology but employs
it to invent a place we can love and care for without loving it to death?

An inventive proposal for so doing comes from my friend Tom Chase
at the Nature Conservancy. He calls it Village and Wilderness, and at its
core are the complementary principles of redevelopment and *un*development. Driven by the collaboration of conservation organizations,
housing organizations, regional and municipal planning agencies, and
local businesses, Chase's Village and Wilderness strategy positions conservation and housing as natural allies whose complementary interests
are rooted in healthy communities that depend upon both sufficient
affordable housing and careful land preservation and restoration.

Here's how the Village and Wilderness plan works:

The Nature Conservancy has conducted sophisticated mapping of
Martha's Vineyard that identifies high- and low-conservation-priority
areas, based on what's necessary to protect biodiversity (including the size

and location of habitats to be protected or restored) and to promote agriculture and aquaculture. The mapping is not constrained by what exists;
it disregards current land uses and the ecological health of the habitat.

Now that the mapping is complete, efforts can be made to rezone low-
priority areas to accommodate clustered housing (the "villages") modeled after traditional New England village townscapes, with sufficient
land to accommodate both residential and commercial activity.

The Nature Conservancy or another private nongovermental organization will raise low-interest loans (and donations to pay the interest on
those loans) to purchase homes in the ecologically important areas (the
"wilderness"). Hopefully, the local land bank will buy these properties
from the Nature Conservancy or other agency through installment payments over ten years. The structures on these properties will be rented
out for ten or twenty years. The proceeds will be used to subsidize affordable housing in the low-priority areas.

At the end of the stipulated time, all structures in the wilderness area
will be moved to village areas or torn down. The land will revert to permanent conservation and restored to native habitat. This is called *undevelopment*. The undevelopment will be repeated as often as necessary to
restore economic and ecological balance and to create enduring patterns
of community and landscape.

The boldness and breadth of such a vision can only be appreciated over
a time span of decades and centuries. Its success requires a cooperative
vision shared by all community stakeholders. This kind of vision is rare.
But it does exist. It's the vision of the cathedral builders. When it is effectively put in place among a community of stakeholders, its power to
shape the economic and cultural life of a region can be monumental.

Mondragon

One of the most successful examples of long-term thinking and community building is the Mondragon cooperative system in the Basque region
of Spain. My visit to Mondragon opened my eyes to the power of economic democracy to create an enduring legacy for a whole network of
businesses and an entire region.

In 1941, a young Jesuit priest named Don Jose Maria Arizmendiarrieta was assigned by the Catholic Church to the Basque town of Mondragon. When he arrived, the area was locked in poverty and still recovering from the devastation of the Spanish Civil War. Don Jose Maria himself had narrowly escaped being put to death for his participation on the Republican side. He believed that part of his service should be to raise the economic fortunes of the people of Mondragon. He founded a technical school. In 1955, five of the graduates, with his assistance, founded a small worker-owned company to manufacture kerosene stoves. More coopera- tive businesses formed during the remainder of the '50s, and Arizmendi encouraged the formation of a bank that, when it opened in 1959, became a cooperative credit union dedicated to the establishment of more cooperatives. The association of cooperatives continued to grow; in the '70s a research institute opened to help with technology develop- ment, and in the '90s Mondragon University, a private university dedi- cated primarily to the study of business and commerce, was established.

Don Jose was a powerful inspiration. He died in 1976, but the Mondragon Corporacion Cooperativa (MCC), which binds together all the cooperative enterprises, continues to thrive.

In January 2001, sixty years after Don Jose's arrival, I visited Mondragon with a small group of Americans for a four-day examination of the culture of both the town and the MCC. Having used a version of the Mondragon principles as the basis for the restructuring of South Mountain Company fourteen years earlier, I found it thrilling to see firsthand the work of this system of cooperatives, which appears to be unparalleled in its dynamism and its impact on a region.

MCC now consists of well over one hundred cooperative businesses, the bank (which has 132 branches in the Basque region), the research institute, and the university. There are over sixty thousand employees. Roughly half of these are owners. MCC's gross revenues in 2001 were over $8 billion, making it the largest corporation in the Basque region and the seventh largest in Spain. MCC's co-ops include Spain's largest producer of refrigerators, leading tool and die makers, and many other industrial companies. They make forklifts, windmills, bicycles, appli- ances, nails, wire, boilers, health and exercise equipment, automobile parts, furniture, woodworking and machine tools, specialized electronic products, manufacturing machinery and robots, and dozens of other industrial products. MCC's Eroski is the largest supermarket chain in Spain, and it does catering, dairy farming, greenhouse horticulture, and rabbit breeding as well. Other co-ops provide engineering, market research, and consulting services. Some develop housing in the area. Mondragon has created a total system wherein one can learn, work, shop, and live within a cooperative environment. The town, in its isolated valley, has a vital, prosperous feel—a small bustling city with a comfortable mix of young people from the university, new middle-class families, and those who have been in the valley for generations. The surrounding hills are verdant and productive, dotted with villages and farms. The MCC's influence reaches into every aspect of community life.

We visited Fagor Electronica, a large producer of domestic appliances. The sprawling factory looked like any other from outside. Inside it was clean and bright. Nobody seemed to be moving too quickly—the atmosphere was relaxed and not too noisy. People worked in small teams on assembly lines, and they traded places with one another from time to time to relieve tedium, and so each could learn the several jobs on the line. A mezzanine level housed offices, meeting rooms, and a coffee bar. The rooms were glassed in and highly visible from the floor. Floor workers freely walked up to conduct business upstairs; managers were comfortable on the floor. There was a sense of camaraderie, teamwork, and integration. We were told that this workplace was highly productive. It felt that way.

Success is an institution at Mondragon. There are several new start-ups each year, and only two have ever failed. In the United States more than 90 percent of business start-ups fail within the first five years. Mondragon is impressive. Legal scholar Peter Pitegoff says, "Mondragon stands out as the most successful coordinated complex of worker cooperative enterprises in the world, with demonstrated capacity for economic growth and long-term survival."[8]

The visit, however, produced many questions. Why don't Mondragon cooperatives produce more for local markets? What drives their recent interest in export and locating new plants overseas? Will their expansion into global markets cause an erosion of fundamental values? Why doesn't their environmental consciousness (which seems higher than normal but not extraordinary) match their concerns about social equity and economic democracy? Can they continue dancing on the tightrope between ongoing success in a global business climate and remaining true to their core values? Pitegoff again:

> While the core enterprises in northern Spain remain cooperatively structured, the MCC has acquired a number of other firms and entered joint ventures in other nations, without implementing worker ownership or democratic governance. Nonetheless, MCC continues to strive for a balance between its democratic social goals and its survival in a competitive domestic and global market. Despite recent expansion, MCC maintains a firm commitment to its cooperative core and sustains a dialogue about extending democratic values to affiliated enterprises. As MCC thrives in a complex global economy and matures as an enterprise, it continues to suggest both the potential and the challenges of worker ownership as an element of democratic economic development.[9]

In response to our questions, our hosts mostly stuck to the pragmatic: Their primary purpose is the creation of safe, secure lifelong jobs and prosperity for their employee-owners. They must do what they do best to achieve this, and sometimes they must compromise in order to achieve this.

My biggest question is this: why is Mondragon such a secret in the United States? It has attracted great attention worldwide, but relatively little here. Even the U.S. socially responsible business movement pays it little mind. Is the idea that capital is a tool, rather than the residence of power, too radical to embrace? Overall, the achievement is splendid. Mondragon turns American-style capitalism on its head. Rather than awarding profit and control to capital, Mondragon has succeeded by awarding profit and control to labor in a system of democratic capitalism. Its cooperatives have found an enduring way to use capital productively and to distribute income equitably at the same time, creating a model of economic democracy that skillfully blends the collaborative efforts of democratic business, collaborating nonprofits, and local government.

In 1941 Don Jose Maria Arizmendiarrieta began to build his "cathedral." Mondragon exemplifies cathedral thinking at its best, and it has grown into an enduring legacy. I wonder whether its current incursions into the global economy are merely digressions in the larger ongoing process of self-invention.

What similar questions will our small company have to grapple with? As the slices of the equity pie become slimmer due to increasing numbers of owner-employees, will the urge to grow become irresistible? Will we find ways, as we build our cathedral, to withstand the pressures and keep our core values intact? Will the cornerstones settle beneath the load?

Handmade spiral. (Photo by Brian Vanden Brink.)

⌐10⌐

THE COMPANY WE KEEP

I'm commonly asked why we are named South Mountain when there are no mountains on Martha's Vineyard.

When we started out in New York State, we worked out of a shop we built on South Mountain Road. Chris carefully handpainted the name on the door of our old flatbed truck. When we got to the Vineyard, we didn't want to bother to repaint, so we kept the name. Years later, I tried to change it, but people in the company reacted so negatively I never tried again. I suppose the name is endearing because it harks back to our irregular beginnings. Perhaps it's also comforting that something—anything—stays the same, given the constantly changing nature of our business.

Tachi Kiuchi and Bill Sherman, in *What We Learned in the Rainforest*, say, "We don't believe change has to seem draconian to be fundamental. It's hard to build a tree but easy to plant a seed. Building a tree is draconian, desperate, and ineffective. Planting a seed is fundamental, serene, and easy."[1]

Having planted the seeds of democracy and craft and cathedral thinking, we can only guess at the shape of the trees that will result. What will the next generation of owners do?

In the first chapter, I wondered whether small business, supported by cornerstone principles like those that have developed at South Mountain, could help make better lives and better communities. I asked whether they could help us be kinder to ourselves and one another, to the planet, and to our children. The arc of this book is an attempt to find answers to these questions by tracking back through the experiences I've had as part of South Mountain Company, and by considering this experience in the context of the work and thinking of others. This process has led to many hunches and conjectures and some new understandings, but no answers yet, as far as I can tell.

Here's what I'm thinking now. I thought, when I started writing, that I wanted to tell a complete story about something we had started, grown into, and become. I see now that the story is not about what South Mountain has become as much as it is about what we have begun. There are things that we have learned to do, that we can do, right now, reasonably well. We can make good houses. We can develop other enterprise that extends our primary endeavors. We can develop effective relationships with those who are connected to our business. We can support a community of employees and employee-owners, giving them the opportunity to make good livings. We can offer, to one another, ownership and a voice in the conduct of our work. We can contribute to the well-being of our community and create small successes that enhance it. We can carefully consider those decisions that affect the evolution of our enterprise. We can infuse our work with a cooperative spirit that feels better than competition. We are beginning to have the capacity to think long term. We're off to a good start. But there is much more to do, and much more to know.

I don't know yet, nor do I know whether I will ever know, to what degree we can build on the foundations we have created and to what degree we can improve our skills. Neither do I know to what extent our experience can help others go down the path toward economic democracy and community entrepreneurship. I don't know whether, in time, many more people will share ownership and control of the companies they work in. I don't know whether local economies will experience a resurgence or whether globalization will continue on its current trajectory, at the expense of bioregional health and local culture.

In *The Soul of Capitalism* William Greider writes:

> The idea of reinventing American capitalism sounds far-fetched. I can report, nevertheless, that many Americans are already at work on the idea in various scattered ways. They are experimenting in localized settings—tinkering with the ways in which the system operates—and are convinced that alternatives are possible, not

utopian schemes but self-interested and practical changes that can serve broader purposes. This approach seems quite remote from the current preoccupations of big politics and big business, but this is where the society's deepest reforms usually have originated in the American past. The future may begin among ordinary people, far distant from established power, who are brave enough to see themselves as pioneers.[2]

I feel hopeful about the effort. Still, I don't know whether my optimism has sound basis or derives from wishful thinking. I would like to know. I would like to be convinced that we will combine our efforts in many places small and large and make a better place for our children. Here is what I think I know.

What has happened at South Mountain during the past few decades has led to some surprising outcomes that, had we followed a different itinerary, we would not have achieved. The fact that we can do some good things with some degree of success and satisfaction is testimony to the spirit of a group of people who have ownership, voice, recourse, and tools. The fact that we, as a group, have the freedom to determine how much or how little we wish to grow empowers us to self-consciously, in the best sense of the word, create the setting for our endeavors.

I think I know that the ongoing attempt to balance multiple bottom lines has been, and continues to be, a worthy, enriching endeavor.

As I write, there's a timber-framing project going on in the annex to the shop. Foreman Billy Dillon, along with Donny Turnell, Dennis Thulin, Kane Bennett, Ken Leuchtenmacher, Mike Marcus (the client, who's working with the crew), and longtime mariner Jon Lange are all working together. They've been at it for a week or more. There is endless banter and a sense of camaraderie.

I check in most days, just to say hello. I notice that the first four are always working in close quarters, together, on the main timbers. Jon is always set up somewhere else, in an outlying location, with his own group of smaller timbers.

One morning I say, "How come you guys never let Jon work in the middle here? How come he's always off on the side somewhere?"

Everyone chimes in at once. "Oh, that's his thing; he chooses it. You know, doesn't wanna hang around with us slobs. Likes to keep out of our madness."

At that moment Jon walks by and mutters, "Around here you learn to hug the edge of the channel."

Hugging the edge of the channel. Steering clear of the crowded middle. Our balancing act—the careful but thorny attempt to harmonize financial success with social progress and environmental responsibility—requires that we hug the edge of the channel. We don't expect smooth sailing, but we do expect to avoid major collisions.

I think I know, too, that our craftsmanship, in all things, is the central thread that makes visible and tangible the underlying principles that guide us. Surrounded by the things we make, we are constantly reminded of the expressions and collaborations from which they resulted. Yvon Chouinard, the founder of Patagonia, the outdoor clothing maker, says his company's goal is simple: everything they make, every piece of clothing or equipment, should be the best in the world. We have the same goal. Recently I got a pair of Patagonia socks for Christmas that wore through the heel in two months, and yet I *still* think that Patagonia actually does what Chouinard says—they make the best. I took those socks back, and the store happily gave me a new pair that won't wear through. Our doors warp, too, and tiles crack, and finishes have blemishes. We fix these so that what we produce is the best it can be. Fortunately for Patagonia and for South Mountain, there's no way to measure whether what we make *really* is the best in the world, but the aspiration is genuine. Craft is a guiding star.

I think I know that we belong here on Martha's Vineyard. Not because it's special. Not because it's different. We belong here simply because we *are* here, because we've been here, because we know that we will stay here. There is virtue, poet Gary Snyder says, in "staying put." That may be enough to know about that for now.

I think I know that our efforts to preserve community through affordable housing are bearing fruit. We may not solve the problem, and we may not win the battle—and it is a battle—but we will make a difference. Bringing stability and security to a single family makes a difference, and that much we have surely done. It's like the story of the kid and her father who are walking on the beach when a particular tide has washed

Detail of driftwood railing with scribed joinery. (Photo by Randi Baird.)

up thousands of starfish. The kid wades into the mass of starfish, picks one up in either hand, trudges to the shore, throws them in, and turns back for more. On the second trip, the father says, "Do you really think you'll make a difference, trying to put them back one by one?"

She holds one up and says, "It's going to make a difference to this one."

I think I know that the idea of community entrepreneurism has taken hold within this company. In 1995, for our twentieth South Mountain anniversary, we had a wonderful celebration with clients, associates, sub-contractors, and friends from the larger community. We made photo albums, Derrill did a series of photos of clients, and we hung the shop and office with mementos and memories. It was good.

Now we're about to celebrate our thirtieth anniversary. At a recent board meeting we began a discussion: What should we do? We talked about what kind of celebration we should have, but nobody got excited. The conversation lacked energy. Then someone suggested that rather than have a party and celebrate ourselves, which would cost time and money, why don't we plow those same resources into doing something wonderful for the island, to celebrate the community that sustains us? We didn't figure out *what* to do, but we did agree that it's *how* we want to celebrate. The discussion was lively. A wind turbine for the high school?

Finally, I think I know, because the tenor is so pervasive, that there is a developing notion of legacy, at least among some in the company, that hints at a bright future.

These are the things that I think I know.

I also recognize, however, that things won't necessarily go the way I wish. Our business fortunes could go sour—the fact that everyone has received a paycheck without fail for thirty years, and that the company has earned a profit every year since we began to keep records, could change in a heartbeat. People could lose hope if things became too hard or took too long or became too divisive along the way. It could turn out that the line I think doesn't exist—the one beyond which all hope is lost—*does*, in fact, exist for our island community, and that we will cross it. It could turn out that competition prevails over cooperation to such a degree that cooperation becomes a soft and sorry road to failure. Any and all of those could happen. Or other unfortunate things I haven't thought of yet.

But let's say none of that happens. These first few decades have been a beginning. Let's say that we can do another few decades, and another few after that, and maybe more. Given all that has happened in just the first few decades, I have to guess that things can get a *lot* better than they are and go a *lot* farther than we've come. If we look at the alternatives to going where we're going, it seems particularly sensible—not visionary, not risky, not wacky, and not trivial, but just sensible—to try to continue down this path. And, perhaps, to take it farther.

Franchising South Mountain

I'm thinking more about franchising. In chapter 3 I talked about the idea of a "knowledge franchise," about franchising what we have learned. I said then that this idea, as it tumbled around in my head, had not fully formed. It still hasn't, but it has begun to take more shape. It may have a place in our future.

In 2003 a group of University of Oregon architecture students were doing a studio project on the Vineyard for their spring term. They visited us. One of the students, after touring our facilities, asked, "If this business and your design/build approach is so successful, why aren't companies doing this all over?"

"Well," I said, "there certainly are some, but it's a good question. I doubt there are many that operate the way we do, but I can't say for sure."

This got me thinking. I can reel off the names of a handful of design/build companies that have some characteristics and principles similar to ours; Wolfworks in Connecticut, Big Timberworks in Montana, New Energyworks in upstate New York, Bensonwood Homes in New Hampshire, Cascade Joinery in Washington, and Graham Contracting outside of Boston come immediately to mind. These companies are remarkable in many ways. I know there must be many others, but here's what I suspect there are far more of: highly skilled builders who don't design, talented designers and architects who don't build, and small residential practitioners of both kinds who are dissatisfied with the work they're doing and disappointed with the fragmentation they experience. Others may be doing well but trying to figure out how to have more meaning in their work, and what to do with the businesses they've built as they (the owners) age.

As Tim Smit, founder of the wildly successful against-all-odds Eden Project in Great Britain, says, "There is a massive hunger in our generation, a feeling that people are not putting their talents to best use."[3]

Maybe we at South Mountain could assemble our expertise in ways that would be useful, in ways that could satisfy some of that hunger. The products and services we could assemble include the following:

- integrated design/build skills, methods, and business systems
- building science, green building, and renewable energy expertise
- affordable housing and community entrepreneurism skills
- amalgamated purchasing of green products to stimulate manufacturers and bring down prices
- owners' manual formats and templates
- employee ownership conversion and long-term legacy assistance
- meeting facilitation and decision-making training
- methods of evaluating corporate social responsibility and balancing profits with service

- strategies for gaining confidence of clients and community
- a coordinated, self-regulating, electronic medium for information exchange, like Great Harvest's

Along with these we could offer a healthy dose of optimism, sup-portive guidance, and the benefit of learning from the many mistakes we've made along the way. It might save others the difficulty of laboring too long before recognizing the obvious, as we often have.

Organizational consultant Robert Leaver says that most people think about power as a pie to be divided—if I give you a piece, I have less left for me. Leaver asserts that power is infinite, however, and that if I give some to you, there is now more of it. Isn't knowledge the same? If I share knowledge with you, we both have it. And if we combine our knowl-edge, new knowledge results. That's precisely what Great Harvest Bread Company does: they share and build knowledge. Here's how you can make the best bread in the world, they say. We'll give you the informa-tion. Then you can do it your way.

In our line of work, there's no question that people must do it their way, because local conditions vary so widely. We can't try to teach you what to design and build, because what works in our area and for our clientele may not work for yours. We could, however, teach processes that could be translated into practice and product.

"Yeah, sure," we sometimes hear. "You can do it on Martha's Vineyard, but . . ."

I am certain that in every college town, resort area, resurgent city, and desirable rural region in this country there are boutique builders and architects and a slew of other types of service-oriented small businesses serving upscale clients and early adopters. The New England states, the West Coast, the Rockies, Minnesota and Wisconsin, the Ozarks, the Carolinas, and the mid-Atlantic states are full of such enterprises. Do these enterprises also serve diverse social purposes? Do they practice and cultivate workplace democracy? Do they have an absolute commitment to their locale? Mostly not, in my experience. Would they like to? I think in many cases the answer is yes, but the culture does not support these approaches.

My thinking is to franchise in the old sense of the word: "to make or

Home—pure and simple. (Photo by Derrill Bazzy.)

set free." The preparation of the information we have in order to make it useful for others would involve substantial work. If we did that work, maybe we could set it free, and perhaps free others to find new ways to enhance their work and enliven their communities.

I don't imagine that South Mountain will ever offer franchises in the true sense of the word. The "franchisees" would be more like part of a loosely amalgamated business network. We could learn from one another. We are at the beginning of something that we think will endure, and we may be just far along enough to share it with others.

We'll see. As I said earlier, you can't count the apples in a seed; you can only plant the seed, cultivate and nurture, and see what grows.

New Job

My career here has had a distinct pattern. I learn how to do something and gain reasonable skills and experience with it, and then someone else comes along who can do it better, while I move on to something else. I was a competent cabinetmaker and furniture maker, but we now have people who can run circles around the highest levels I was ever able to achieve. The same goes for carpentry; my skills and talents have been surpassed by orders of magnitude. I used to design the houses we built and spent years learning to do a credible job. Now others do that. I used to be the person who found the oddball tiles, old stained glass, or soulful furnishings that would make a project special. Now Deirdre, who runs our interior design department, is out on the prowl turning up new stuff, working with artisans, and enlivening our buildings in ways that go far beyond what I did. For many years I have overseen all aspects of the business; my sabbatical meant others had to take on more management responsibility, and they do it well.

Now, having returned, I find that my job has begun to change again. As always, it's not without sadness. But from each endeavor some part stays with me and informs the new things I do. The great reward is that I get to watch and help others do these things I care about, and I get to see them done with skill, enthusiasm, and concentration.

According to the history, those parts of my job that others are now moving into will be handled better and better. Others will grow into the areas that I am growing into now, and eventually beyond. It will happen in due time, as we're ready and able. It will happen at a pace that works for us. It will not happen without bumps in the road. We continue to be a work in progress, a foundation on which we layer our accumulated experience. We sustain the journey, always practicing and adjusting, tearing and mending, folding and unfolding, building the road as we travel, shaping our future. This company's failures and weaknesses stand side by side with its successes and fulfillments.

It's the company we nourish, the company we test and challenge, the company we hope will endure and continue to enjoy the opportunities conferred upon us by this place. This is the company whose care is entrusted to us and whose success requires our relentless dedication. This is the company we will keep.

Right now, I feel like I'm standing on deck, looking over the rail, watching the water as our ship cuts through. I am enjoying this moment in time. I am humbled by the generosity of all those who have shared the journey, from this morning's discussion at coffee break back to the early

Looking through an old gnarly at a house on the North Shore. (Photo by Brian Vanden Brink.)

days at the Allen Farm when Mitchell and I were building my parents' house, and farther back to the image I hold of my grandfather steaming into New York's harbor in 1899. The observations and stories I've offered tell who and what we are now, as it appears to me. Most likely others who have been in the thick of it would tell it differently.

If the discoveries from the beginnings of our journey inform yours, I hope you will share your findings and further enlighten ours. Together, perhaps, we can share the *very best* of what we learn, and the love that comes from the learning, with our children, and theirs, as they invent a future we can't even imagine.

ACKNOWLEDGMENTS

We think of writing a book as a solitary enterprise. The author, alone with a collection of ideas, tries to convey them to others in a compelling way. The experience of making *this* book, at least, has been different from that, and as highly collaborative, in a way, as anything I have ever done. No part was done alone.

The book is a product of countless shared experiences. The people mentioned in the text, and hundreds of others who are not, have contributed to the undertaking. These acknowledgments are my opportunity to thank some of them—the ones who have specifically, and generously, helped me with this project. But a host of others have contributed in ways they probably do not know and in ways I cannot fully identify. This so-called solitary enterprise seems more to me like a shared journey, akin to building a house.

The journey has been important to me as part of my work, but as part of my life, too. And nothing is more important to me in life than my family—my wife Chris, my kids Pinto and Sophie, my grandchildren Kalib and Silas, and their mothers, Kristen and Jessica—they bring me unremitting joy.

My parents, Marilyn and Herb Abrams, have supported me in so many ways they can't be counted or measured or even, I'm sure, fully recalled. They have been rock solid, always there for me, and my father's work ethic, optimism, and sense of commitment have been instrumental in shaping my own. And my sister Nancy helped me out of the starting blocks: each day she taught me everything she learned in school, when I was three.

I treasure them all.

Three people have been essential mentors to me in writing this book and more: Jamie Wolf, Lee Halprin, and my wife, Chris Hudson Abrams. I have been engaged in a discussion with Jamie about business and

community for a more than a decade. It started when we met as co-conspirators at the Northeast Sustainable Energy Association in the late
1980s. We organized conferences together, struggled to help build a good
nonprofit into something even better, and enjoyed each other's company.
Jamie runs a design/build remodeling company, and we came to find that
we shared many ideas about business . . . and life. Our ongoing discussion
has been mostly by e-mail. Every so often we get tired of the electronic
back-and-forth and pick up the phone or find a way to meet. One way or
another the discussion goes on. It always challenges me to think harder.
Jamie has been an avid, critical, and supportive reader as I've been
writing. I'm grateful for the company and the help. I look forward to continuing this inquiry and this friendship together.

Lee goes back a long way in my life, to the beginning of our time on
the Vineyard. He is a relentlessly caring friend and mentor who never lets
me get away with anything if he can help it. Over the years, at regular
intervals, I have had enough nerve to give writings to Lee for review.
They come back smothered in red. Sometimes there is a note at the end
as long as the piece itself. A part of one of those notes is the quotation I
used at the end of the first chapter, about the slightness of our knowledge. A superb editor who probes and queries, suggests and soothes, digs
way deep, and never minces words, he has helped me learn to write a
little, as he has helped me with so many other things.

When I was in the early stages of this project I gave him my book proposal to read. Then, after I had written a few chapters, I passed them to
him and he read them, as always, with care. He made me think differently, in so many ways, and changed the course of this project. If—and I
know this is a big if—knowing what I don't know has somehow seeped
into my writing to some small degree, it is because of Lee.

Along with teaching me more about life and humanity and love than
anyone else, Chris is one of the best bullshit detectors on the planet. It's a
good thing, I think, for each of us to have one of these in our lives. She has
carefully read my writing and looked for those parts that don't ring true. If
you find a lot of b.s. in this book, some small part may be because Chris
missed it, but more likely it's because I didn't listen. But she always does.

She also managed to put up with my absolute self-absorption during
periods of time when I was consumed by my writing. We've been

together since 1969, when we were just twenty years old, so we've been on this entire journey together. In a way, we grew up together. If not for her, I doubt any of this would have happened the way it did, or even at all.

The generosity, wisdom, and goodness of these three people just knocks me out.

There are other people who have been essential to this project. Several stand out.

I met Bill Greider when we were both speaking at the Vermont Employee Ownership Center annual conference. At the bar that night, swapping tales, I had the sense that I was in the presence of a great raconteur whose wisdom was legion. He told me that night that he had had "a low-grade obsession with employee ownership for the past twenty-five years." I liked that. We talked a few times after that. When I crossed paths with him at a meeting in Boston last May, I asked him if he would consider writing the foreword to this book. He graciously agreed. Later, when the manuscript was complete, I sent it to him. He read it and wrote me a remarkably thoughtful letter. Along with plenty of encouraging words, it contained a suggestion for how to make the book better—one single, clear, but large idea. I took him up on it and restructured several sections. It's a better book because of him.

I was skeptical when I learned that Woody Tasch, the chairman of the board of Chelsea Green, wanted to be the developmental editor for this book. He had never edited a book before, and, besides, I know him to be a very busy guy. Would he be good at it? Would he be available? I agreed to give it a try. He has been tremendously valuable: teasing out meanings, analyzing structures, and offering alternatives with great skill. He once substituted the four-letter word *lieu* for eleven words of my prose without changing the meaning of the sentence. I'm grateful to Woody for being there. He did a splendid job. It has been a fine collaboration.

Kevin Ireton, the editor-in-chief of *Fine Homebuilding*, started helping with this project ten years ago, when I wrote an essay for his magazine. At the time I didn't know what an editor actually did. He taught me, by doing so, that an editor has only one objective: to help you say what you want to say, better than you can say it yourself. Kevin is a great editor, a clear thinker, and a great friend.

Mike Drezner is one of my co-owners. I wanted one person in the company to read the manuscript before it was published. He was my choice, because for twenty years Mike has been a piece of the bedrock of this company. He plays many significant roles. He has been a particularly good partner for me. We're very different from each other in ways that are complementary. We agree about a lot. We disagree about a lot. When we disagree, we learn from each other. I also want to give a special nod to two other co-owners, Deirdre Bohan and Phil Forest, who stepped forward and became, with Michael, the executive committee when I took my sabbatical. They did a tremendous job, and they are the future of this company.

Nick Weinstock, an accomplished author and the son of wonderful longtime clients, helped me create the book proposal. Somehow he generously fit my stuff into his remarkably busy life. He helped me understand that I could do this.

I sent my proposal to only one publisher, Chelsea Green. I had an intuitive sense that Chelsea Green had commonalities with South Mountain, and that their values foreshadowed a successful collaboration. Margo Baldwin, Chelsea Green's publisher, accepted the proposal, apparently without hesitation. She called and said, "These are the precise issues we are grappling with in our company. I want the book." I'm grateful for her confidence—she's been with it all the way from the start, and I look forward to a long association with Margo and Chelsea Green.

Marcy Brant has been the editor at Chelsea Green saddled with the responsibility of overseeing and shepherding this project, and dealing with me. She has done so with huge grace and skill. She misses nothing. She and the others at Chelsea Green, including but not limited to Collette Leonard, Kelly Manning, and Erin Hanrahan, are a joy to work with. My frequent questions—as a rank amateur—must have been annoying, but I never got that sense. I am in very good hands at Chelsea Green.

Peter Holm and his assistant Daria Hoak, at Sterling Hill Productions, have done a superb job with something very important to me—the design of this book. Nancy Ringer did a great job with the copyediting—she really got it, and got me, and made insightful suggestions.

Just before I sent the proposal to Chelsea Green, I sent it to three agents. One of them, Upton Brady, responded positively, but his response came after I had decided to send it to Chelsea Green, and after Margo had called and indicated that she wanted the book. Upton agreed to take it, also, and in his letter he said he thought it would be perfect for a small, independent publisher in Vermont: Chelsea Green! I had already agreed to go with Chelsea Green, so I had no need of an agent, but Upton was very helpful in several ways, and I appreciated his confidence.

Other people read parts of the manuscript along the way and were immensely helpful: my father, Herb Abrams, and friends Jonathan Orpin, Nina Keller, Jeff Halprin, and Carol Evans.

I also wish to thank all the photographers who took photos that appear in this book, and I especially want to mention Betsy Smith, my office assistant here at South Mountain, for the tremendous help she has been to me in this project—organizing and managing the photos, transcribing my endless underlinings from books and magazines, doing the bibliography—you name it, she did it. She's the best.

The real heroes in this story are the people of South Mountain—my colleagues, co-owners, and coworkers. I am deeply appreciative of all that each of them has been and done. I want to say each name, because each person has different meaning to the company, and to me personally. First, my co-owners: Pinto Abrams (my son), Derrill Bazzy, Kane Bennett, Deirdre Bohan, Peter D'Angelo, Bill Dillon, Mike Drezner, Phil Forest, Pete Ives, Ken Leuchtenmacher, Peg MacKenzie, Tim Mathiesen, Peter Rodegast, Jim Vercruysse, and Laurel Wilkinson. Because of all these wonderful partners I was able to take the sabbaticals I took, and I was able to write this book. It's amazing to me to think that one of them, Pete Ives, has been here twenty-eight years, just two fewer than I have! And he's still the dean of Vineyard surfers.

Next, the other South Mountain employees and, hopefully, future owners: Ryan Bushey, Ben Cameron, Curtis Friedman, Bob Julier, Jon Lange, Rob Meyers, Siobhan Casey Mullin, Steve O'Brien, Greg Small, Betsy Smith, and Don-E Turnell. And the sometimes employees too: Valerie Reese, Tara Simmons, Dennis Thulin, and Jill Walsh.

To the two former owners, Steve Sinnett and Vicki Romanauskas (now Sperry), I want to say thanks for paving the way. From the beginning, Steve has been immensely important as both a friend and a coworker.

I want to thank all our former employees, too: Mary Alpern, Eric Bates, Lou Botta, Isaac Canney, Marc Carroll, Meredith Dillon, Steve Donovan, Woody Douglas, Tim Eddy, Greg Hise, Bruce Ignacio, Patrick Lindsay,

Primo Lombardi, Scott Mullin, Dana Petersen, Carl Pratt, Carlos Ramirez, Nancy Rodgers, Eric Ropke, Heikki Soikkeli, Alicia Spence, Phil Strother, Marco Turoff, and Suzanne Williamson. I include even those few we parted with uncomfortably because I'm certain I could have done something to make it go better.

And the kids who have worked summers over the years—some have gone on to great things, while others are still going there. They are Sophie Abrams (my daughter), Vamp Campbell, Nat Cohen, Mark Couet, Milo D'Antonio, Darin Evans, Sean George, Taylor Ives, Insley Julier, Tim Laursen, Scott Lazes, Chuck Leger, Eben Light, Wesley Look, Sam Miller, Skye Morse, Jack Reynolds, Danny Sagan, Josh Vag, and Noah Yaffe. Danny started when he was fourteen, in 1980, when we built his parents' house. He worked for many years after and he's now an architect and design/builder in Vermont. Noah worked with us for years, too, and he also went on to architecture school and recently finished.

There are scores of others who have played essential roles in my learning. I'll single out a few—some from long ago, some more recent, and some who were gracious enough to be interviewed (and probably misrepresented) for this book: Merle Adams, Clarissa Allen, Madeline Blakeley, Dick Bluestein, Anne Boothe and Toni Bishop, Stewart Brand, Terry Brennan, Phil Brougham, Tom Chase, Bruce Coldham, Mike Ferretti, Smokey Fuller, Greg Graham, Lorie and Richard Hamermesh, Terry and Jerry Hass, Fritz Hewitt, Marjorie Kelly, Rob Kendall, Matthew Kiefer, Jerry and Nancy Kohlberg, Bob Kuehn, Richard Leonard, Ed Levin, Tony Lewis and Margie Marshall, Brian and Anne Mazar, Gino Mazzaferro, Dennis and Nancy McHone, Peter Pitegoff, Christina Platt, Mitchell Posin, Job Potter, Ron Rappaport, the guys at Red House Builders, Marc Rosenbaum, Eli and Frimi Sagan, David and Pat Squire, Jerry Tulis, Roy and Diana Vagelos, Juleann VanBelle, Kingsley VanWagner, Kate Warner, Davis and Betsy Weinstock, Allen White, Alex Wilson, my cohousing co-conspirators Paul and Sylvie Farrington and Philippe Jordi and Randi Baird, all the folks at Indigo Farm, and the many subcontractors, suppliers, and associates of all kinds who have been such important parts of our successes through the years. And finally my in-laws, Martha McGuffie and Perry Hudson, who helped us get started back on South Mountain Road.

Now what? Just sit here and try to think whom I may have forgotten? That list would likely be even longer.

You can contact me at jabrams@vineyard.net. The South Mountain Web site is at www.southmountain.com.

APPENDIX ONE

South Mountain Employee Ownership Particulars

South Mountain Company, a sole proprietorship, restructured in 1987 and became an employee-owned cooperative corporation under Massachusetts law. I sold the company to a group that consisted of myself and two others. We became three equal owners of the new corporation.

My compensation came as a full ownership share in addition to preferred stock, which, together, equaled the valuation of the business at the time. The preferred stock was a onetime device to compensate me for the sale of the business, and was converted to cash over a period of five years. The first board of directors meeting of the newly reorganized company convened on January 9, 1987. Within the new structure we successfully expanded both the ownership and the business itself. We began with three owners in 1987; after losing one, we gained seven more, for a total of nine, and then lost another in July 1995. Since then we have gained another eight, for a current total of sixteen employee-owners. There are thirteen other employees who are currently on a track toward ownership and one who has decided not to become an owner.

This is, as my friend Jamie Wolf says, "a look under the hood" of the structure of South Mountain.

Ownership Criteria and Eligibility

All employees are considered to be prospective owners. To qualify for ownership, employees are expected to meet four essential criteria:

1. The intention to be employed by South Mountain for the foreseeable future.
2. An ability to work effectively and cooperatively. Evaluations should demonstrate steady improvement where necessary, the ability to accept criticism, and the desire for self-knowledge.

3. A commitment to understanding and honoring the company's core
   values: quality work, ethical business conduct, environmental responsibility, and concern for other people; in other words, we expect that a
   new owner will be a good representative of the company.
4. A commitment to having South Mountain be one's primary work.

Ownership eligibility begins once an individual has worked a minimum of five years and seventy-five hundred hours. Employees are evaluated for ownership suitability and educated about the meaning of
ownership during their first five years of employment. This process is
conducted by our personnel committee. The intention is that when an
individual reaches eligibility, it will be clear whether the individual wishes
to accept the responsibility and whether the current owners wish to
accept the individual as a new owner. It is not always quite that simple,
but the process is generally orderly and consistent.

Governance and Management

Decisions are made by the board of directors, which consists of the company owners. All owners are employees; there are no outside owners or
board members. Each board member has one vote, but we work by consensus unless we are unable to achieve it. We are committed to the consensus decision-making process. In seventeen years we have had to take
only three votes. In the first case a rather minor issue provoked strong
feelings. We couldn't reach consensus, and finally someone said, "We're
spending way too much time on this meaningless issue—let's take a
vote."

The second and third votes were more substantive. The second was
about an important and complex proposal to buy property, relocate, and
consolidate the business. There was disagreement about the merits of
the proposal, and some board members felt that the process had not been
conducted in the way all would have wished. We had three long meetings
about the subject in one week. We couldn't come to full agreement, so
we took a vote. The proposal passed easily.

The last vote was about a prospective new owner. It was marked by strong disagreement and debate. The proposal was passed, but the experience spurred us to revisit our ownership criteria and clarify our policies. These exercises proved to be valuable. We wished we could have reached consensus in all cases, but we were grateful for the voting mechanism. It's an essential backup when needed.

The board has responsibility for decisions affecting the future of the company, such as the following:

- accepting new owners and other significant personnel issues
- compensation and benefits policies
- profit sharing
- direction of future projects and work
- major purchases, investments, and expansions
- new ventures
- company growth
- involvement in community projects
- major donations

Management prepares the board to make these decisions and in most cases recommends direction and courses of action for the board's consideration. Before each monthly meeting an agenda and a package of supporting material is distributed to prepare the board for discussion and deliberation.

As the owner group has grown we have begun to do more of our work in smaller committees. An executive committee of four owners meets each week. There are three other active board committees: the personnel committee, the charitable contribution committee, and the education committee.

There are five distinct areas of the company, each consisting of a team of employees: business office, design, interiors, woodworking shop, and carpentry crew (there are currently three crews). With company growth we have developed more collaborative management processes to go along with our shared policymaking. My job is to manage the company, but much of our management work is done by committees, and gradually it is becoming more and more the responsibility of the executive committee to manage the company. A production committee and a design committee meet on a regular basis. A small committee manages our facility. There also are individual management roles; each job is managed by a foreman, and there is a shop manager, a small-jobs manager, a manager of the interiors department, a product sales manager, and an information technology manager.

Ownership Responsibilities and Benefits

Responsibilities

1. *Payment of ownership fee.* See "New Owner Payments," below.
2. *Attendance at and participation in board meetings.* Owners are expected to digest material in board packets and to attend all board meetings. This requires an understanding of South Mountain operations and mission.
3. *Understanding of South Mountain governance.* An owner should be familiar with South Mountain bylaws and should know how the governance and internal capital accounts work. A new owner must learn what it means to act in the best interests of the company.
4. *Representation of South Mountain.* In a way, each owner is a community ambassador for South Mountain. We expect that each of us will conduct ourselves in ways that are consistent with the values of the company, as expressed in our bylaws, our mission statement, and our guiding principles.

Benefits

1. *One voice (or vote) on policy matters and one full ownership share.* Ownership is an opportunity, as well as a responsibility, to impact the decisions that chart the direction and destiny of South Mountain, and that determine the quality of each individual's work life.
2. *Equity sharing.* All owners share equity in the form of internal capital accounts. See "Internal Capital Accounts" below.
3. *Ownership position.* This recognition is an intangible that may mean more to some than to others. Owners don't just work at South Mountain—we own it!

New Owner Payments

The payment an employee must make upon accepting ownership status, called the membership fee, is for the purchase of a full share of South Mountain. When we first restructured the company as a worker-owned cooperative corporation we had decided that this payment needed to be significant but affordable. If it was too steep it would discourage participation, so we set it at the price of a good used car, an expense everyone seems to be able to manage when necessary. In 1987 the fee was set at $3,500. Originally we agreed that it would escalate at the rate of 10 percent a year, but in 1993 we agreed to reduce that rate to 2 percent a year. In November 2004 the fee was $11,600. At this point it's an uncommonly good investment for new owners.

When the fee is paid, it is deposited into the South Mountain cash reserve fund. At the same time, the new owner's individual capital account is opened. The fee may be paid (on May 1 or November 1) in cash in one lump sum, or payments may be spread, at no interest, over a period of time not to exceed thirty-six months. A new owner paying over time begins to accumulate equity when the fee is half paid.

Internal Capital Accounts

An important part of the South Mountain system is the building of owners' personal equity through annual profit sharing. Equity is accounted for in the owners' individual internal capital accounts. These are paper accounts backed up by the company's net worth, not cash accounts.

Equity accounts begin with the membership fee that each owner pays and grow at the end of each profitable year. They do not accrue interest. Fifty percent of the annual net profit is distributed among the owners, based on the hours they worked during that calendar year. (This is distinct from annual cash profit sharing, in which roughly 35 percent of the profits is divided among all employees in the form of wage bonuses.)

The board may, however, distribute dividends in any given year in lieu of all or part of the funds that would otherwise go into the internal capital accounts. According to the Internal Revenue Service, at least 20 percent of dividends must be cash, and the entire dividend, even the noncash portion, is taxable income for each individual. The board distributes dividends only in high-profit years when doing so creates a tax advantage to the company, and the board makes certain that the cash portion is enough to at least cover the increased income tax liability generated from the cash distribution, so as to not cause a financial hardship to individual owners.

The equity accounts continue to mature until the end of an individual's employment. Except for dividends, these accounts are generally inaccessible until an individual leaves the employ of the company. There is a provision for limited withdrawals on the following basis:

- Up to $30,000 total per year can be distributed on a first-come, first-serve basis, with a maximum withdrawal of $15,000 per person.
- Distribution is at 60 percent of value (the recipient can take a tax deduction on the remainder).
- The withdrawals are not intended to be used for investment purposes.

To date nobody has elected to withdraw funds.

Upon departure, each former owner begins to receive the equity that is due. Payments are spread over ten years. A departing owner can elect an accelerated payout; if so, the distribution is at a discounted rate as follows:

- Immediate payout: 60.0 percent
- Three-year payout: 72.5 percent
- Five-year payout: 79.5 percent
- Seven-year payout: 86.5 percent

The only payout to date was for an owner who departed in 1995. It was approximately $40,000 (after four and a half years of ownership). At the end of fiscal 2004, the cumulative total of the capital accounts was $1,050,000. Individual accounts ranged from $21,000 to $179,000.

In 1999 we established a reserve fund to back up the equity and to use to pay off owners as they leave in the future. We build this account aggressively. The funds are invested in socially responsible venues; as of December 2004 it contained approximately $600,000. If the owners were to choose to, they could use reserves for other purposes, such as paying employees during a down business cycle. We borrow liberally from it (we use uninvested funds as a no-interest internal line of credit), but as yet nothing has been spent from this fund. It is the foundation of the stake that each of us has been earning.

At present, according to our bylaws, nobody may maintain ownership and share profits beyond the end of employment. At termination or retirement, an owners' share must be sold back to the corporation. We are in the process of changing this provision so that retiring owners could retain ownership in certain situations.

APPENDIX TWO

Meeting Facilitation and Consensus Decision Making

Good meetings are a joy. Poor meetings are tragic. Many of us, in our lives, spend significant time in meetings, but we are not taught how to lead them well or how to effectively participate in them. The prevailing method for conducting meetings, Robert's Rules of Order, comes from military beginnings and relies on rigid structure, rules of conduct, and strict adherence to the rule of the majority. Often nearly half the people at a meeting disagree with a decision that has been reached. In many cases, by using a more open process that encourages dialogue and participation, we can arrive at decisions that are supported, at least to some degree, by everyone affected.

I have facilitated hundreds of meetings. This doesn't mean I'm particularly good at it or have special expertise, but my experience has given me great respect for the practice. I work in a company that operates by consensus, I live in a cohousing neighborhood that is governed by consensus, and I chair a nonprofit that operates by consensus. I find consensus decision making to be an effective, just, and highly collaborative way to make decisions, develop initiatives, and solve problems.

It's clear to me that a workplace is a better place when employees truly work in teams, but again, the most familiar team models we have are those that are created to win wars and games. We have a commander or a coach who gives orders, and the soldiers or the players use those instructions to defeat the opponent. Mediator Bill Ury says, "People are realizing that adversarial, win–lose attitudes in an increasingly interdependent world, where I depend on you and you depend on me, just don't work anymore. Using those tactics is like asking, 'Who's winning this marriage?'"

Who's winning this company? Wrong question.

Facilitation and consensus are two powerful tools for building nonhierarchical teams that can produce the best possible collaborative thinking. I am not suggesting leaderless teams and open-ended processes with no controls. Quite the opposite. I'm suggesting well-led processes that invite, engage, and expand capability.

To talk about facilitation and consensus we need to know what they are. I'll take them one at a time.

Facilitation

To create as a team we must meet as a team. Too many meetings are unsuccessful: too long, unfocused, petty, unproductive, led by domineering or disagreeable individuals, meandering, time wasting, boring, contentious. Such meetings leave participants dreading the next one.

A facilitator is an individual with a particular skill who accepts responsibility for helping the group move through an agenda in the time available and make necessary decisions and plans for implementation in order to accomplish common goals.

Usually meetings are facilitated by default, by whoever is in charge, without recognition of the meaning of the job.

Facilitation is part art and part science. There is a growing body of knowledge and resources about the practice, and there are several ways to learn the skill. It is being consciously practiced more and more in business, government, and the nonprofit world. There are also many natural facilitators—leaders who don't think about facilitation but practice it as their standard mode of conducting meetings.

Especially with large groups, facilitating can be scary—it's a big responsibility. It requires moment-to-moment awareness, being awake and active, and thinking on your feet. It's like dancing. If your mind wanders, you lose the rhythm and stumble. It's more like jazz than classical music, and it's exhausting.

And, just as musicians have off nights, facilitators can have off meetings. I've had plenty.

The responsibility of the facilitator is to the group and its work rather than to the individuals within the group. The group gives the facilitator additional rights to accompany the increased responsibility.

A facilitator encourages the expression of various viewpoints—the more important the decision, the more important it is to have all pertinent information (facts, feelings, and opinions) on the table.

A facilitator keeps the group focused on the agenda item and task at hand.

Perhaps the most common misconception about facilitation is that the facilitator is supposed to make sure that everyone gets to say what they want. That is the opposite extreme of not letting anyone speak at all. Facilitation is the art of finding the middle ground—the place where as many people as possible get to express themselves as completely as possible within the bounds of what the group is willing to listen to and what time permits.

Facilitation is not about encounter. It is about learning a form of group guidance that produces accommodation.

The facilitator is responsible for protecting ideas and individuals from attack, suggesting processes for following the agenda, and devising other approaches if the process bogs down.

Listening is the primary skill of facilitation. The facilitator listens for the whole group and each person in it. As facilitator, you listen for the group's purpose, and the power of your listening focuses and energizes the group.

I have conducted facilitation training at both South Mountain Company and Island Cohousing, and in both organizations we now have a number of skilled facilitators. I have noticed that once they are trained to facilitate, people become better meeting *participants* as well as better meeting *leaders.* They understand the process in a new and more complete way.

Consensus

In my view, the highest goal of facilitation is to produce consensus.

Consensus is a process of synthesizing the wisdom of all participants into the best decision possible at the time. It is not unanimous agreement, and in fact, participants may consent to a decision that they disagree with, but that they recognize meets the needs of the group or the situation. The root of *consensus* is *consent*, which means to give permission to. When you consent to a decision, you are giving your permission for the group to go ahead with the decision.

Consensus is about accommodation, but, more important, it's about nobody having to accept that to which they are *vehemently* opposed.

The cooperative nature of consensus yields a different mind-set from the competitive nature of majority voting. Key attributes of successful participation include humility, willingness to listen to others and see their perspectives, and willingness to share ideas without insisting they are the best ones.

Some describe consensus as a transformational process. When we use the accumulation of several peoples' ideas and weld them together, the final product is better than what anyone could have devised on his or her own. The idea of consensus is not to eliminate conflict but to transform it.

At South Mountain Company we have used consensus decision making for seventeen years to run our business. At Island Cohousing, where I live, we have used consensus decision making for four years of development and five years of living. As the chair of the Island Affordable Housing Fund, and in many other facilitation situations, I use the consensus process even when it is not explicitly stated that we are doing so.

How Does Consensus Decision Making Work?

Consensus can be divided into five parts or stages:

1. Expression of an initial idea;
2. Discussion of the idea;
3. Synthesis of reactions and creation of a proposal;
4. Testing of the proposal within the group, and modification
   if necessary; and
5. Implementation and evaluation of the decision.

The fundamental difference between consensus and majority vote is that in a consensus process a single person can block a decision. Consensus empowers each individual in a way that majority voting does not. Majority voting can accomplish decision making quickly, but it also can strain relationships and the sense of community. In achieving a majority of votes, expediency can become more important than relationship. What one individual thinks may not matter unless that individual has sufficient power. Consensus often requires more creativity, and it often results in more complete solutions.

Because consensus can become paralyzed by one difficult, powerful, or dysfunctional individual, I advocate a backup voting mechanism to be used when consensus cannot be reached after a specified amount of discussion. In the organizations with which I am most familiar, this mechanism has been essential but rarely used. Aside from its practical utility, its existence assures more adherence to the consensus process—when someone is being stubbornly disagreeable, that individual knows that he or she is likely to be outvoted if he or she doesn't find a way to compromise.

Occasions do arise in which individuals are consistently argumentative for the sake of argument. They often characterize their behavior as "playing the devil's advocate." I once heard a facilitator respond to someone who was "just being the devil's advocate" as follows: "Thanks for your sentiments, but I think the devil has all the help he needs."

Consensus is a conservative process. Because it takes a new consensus to replace an existing decision, decisions tend to stand once made. Some people are uncomfortable with this conservatism because it can be hard to change a decision. To address this, some consensus proposals include a review period or a sunset clause. Requiring that the decision be renewed after some time has passed can encourage a group to experiment with new ideas without fear of being locked into a risky or unfamiliar path. It also provides an easy mechanism for incorporating new learning, over time.

One way to ensure that group time is not spent reconsidering previously made decisions when only one person—or a few—wants to do so is to require that reopening a consensus decision have a minimum number of supporters, say 10 or 20 percent of the group.

There are some issues for which consensus may not be an effective process. A classic example is style issues or color or design choices. Choosing the color scheme for corporate headquarters may not be the best decision to put to a group consensus process, because there is no best choice between blue or green; they are simply personal preferences. In these cases, using a weighted voting system on a number of choices may be a more effective way to get the job done.

Degrees of Consent

Consent does not mean agreement. The goal of consensus is to come to a decision that everyone will give permission to, at least for a while. Supporters of a decision usually include true supporters of that position, those who don't really care either way, and those who don't fully support the position but don't wish to stand in the way.

Blocking is appropriate only if a participant strongly believes that a proposed decision is going to be bad for the whole group or to violate the mission of the group. If a participant blocks a group decision because of his or her personal values, that individual is essentially demanding that the whole group subscribe to his or her values. It is the facilitator's job to be clear about this and to remind participants of the powerful responsibilities that come with the ability to block decisions.

There are ways of objecting to a proposal without blocking consensus:

- Nonsupport—I don't agree with this decision but I will go along with it.
- Reservations—I think this decision is a mistake because
  _____, but I'll live with it.
- Call for a later review—I would like this decision reviewed after
  _____.

Decision-Making Reflections

I am sometimes asked whether it is perilous for the employees to make the decisions for a business. What do they know? Isn't it inefficient and potentially paralyzing for decisions to be made by consensus by a diverse group? Shouldn't we leave the decision making to skilled management?

I speak primarily from my particular experience. South Mountain's governance system is a democracy with clear divisions of responsibility and authority. Much of the authority to act is delegated to management. This delegation comes easily, because this was the established mode of operation before the ownership of the company was shared. The difference is that there is now a clear mechanism for discussion, debate, and change. The comfortable delegation of authority may be one of the advantages of a company converting to worker ownership and control, and consensus decision making, rather than starting that way. Once the entrepreneurial leap of starting a new business has been achieved, adoption of consensus-based decision making becomes a part of the maturation process. In our case, consensus decision making has only broadened our view; it has not watered down our decisions or derailed our ability to make them in any discernible way.

To our own I can add the experience of one of the largest companies in the world, Shell Oil, which finds consensus to be an effective form of decision making, as author Arie de Geus explains in *The Living Company*:

> From the top of the Shell Group down there is no traditional mechanism to resolve conflict. . . . One way or another, the members of the Committee of Managing Directors (CMD) and the two boards of directors have to agree among themselves on solutions that are acceptable to all. In practice, there . . . is no good way to force through a decision to which one or more of the members are actively opposed. The minimum that is required is a *quasi-unanimity*. . . . Quasi-unanimity does not mean that everybody agrees with the proposal. It means that no one is so violently opposed that he will show a veto card. The chairman has no other power than his persuasion; he has no casting vote or final decision.[1]

De Geus goes on to say that distribution of power can be frustrating, but it brings more actively engaged minds into the decision-making process. It may lead to better action and more complete organizational learning because implementation is an integral part of the decision, rather than something separate that is imposed later. The people whose cooperation is necessary are right in the thick of it.

Over the years, as we came to recognize and define South Mountain values, we were also learning more about decision making. Gradually we were gaining understanding of the fact that ideas placed before us were meant to be discussed, shaped further, and accepted or rejected as a group. Each was fair game for adjustment, change, back-burnering, or scrapping, so there was no need to fight for or against new proposals; we had only to consider, to listen, to muse, to articulate. Dialogue produced satisfaction; argument did not. Each of us must learn this differently. As the most entrepreneurial of the group and the former lone decision maker, I must learn not to push too hard. Others must learn not to cast themselves as powerless victims. We all have to learn to trust one another. As we continue to work at these things, we have begun to enjoy the conversation, I think, and in-your-corner persuasion has given way to collaborative examination. We can now effectively take a stand for what we believe in, as individuals who are members of a group of owners, as a group that is the steward of the company, and as a company that is part of a larger community.

I don't pretend to cast any of these gains as done deals. We will continue to struggle with these issues of responsibility, accountability, quality, fear, ownership, and direction.

Although both South Mountain Company and Island Cohousing are democratic organizations that rely on participation and consensus decision making, there are obvious differences: South Mountain is a business and Island Cohousing a neighborhood; South Mountain has history and Island Cohousing is relatively new. But there is another difference, less obvious but fundamental: South Mountain is a culture of expertise; decisions are made and actions taken by those who have the ability and experience. Others recognize that and place their trust in that approach, even though they may have the power to do otherwise. There's a strong sense of responsibility to essentials like clients, the bottom line, and company reputation. Island Cohousing, on the other hand, is a neighborhood, and neighborhoods are not about expertise; they're about neighborliness. At Island Cohousing I hope that we are gradually developing a culture of respect for the neighborhood itself, as well as for each other. Evolving traditions and standards and a sense of neighborhood history will, I hope, drive this cultural evolution. It has a long way to go.

One of the truly rewarding aspects of my work has been the long-term process of learning about and practicing facilitation and consensus decision making. Years of experience lead to confidence. When I begin a meeting these days, I generally know that we will reach agreement as needed. The exciting part is what I don't know—what the agreement(s) will be. Surprising conclusions and dramatic divergences from starting points are common when a community of interest engages in a well-conducted process of collaborative decision making.

Facilitation and Consensus Resources

Many resources that consider facilitation and consensus decision making in greater depth are available. I list only a few below. Much of the information in this appendix originally came from the work of Rob Sandelin, whose teachings I have found to be particularly useful and instructive.

- *A Facilitator's Guide to Effective Consensus Meetings*, by Rob Sandelin (available at www.ic.org/nica/book/Cover.htm)
- *What Is Consensus?* by Rob Sandelin (available at www.ic.org/nica/Process/Consensusbasics.htm)
- *Getting to Yes*, by Roger Fisher, William Ury, and Bruce Patton (Penguin Books, 1991)
- *The Third Side*, by William Ury (Penguin Books, 1999)
- *The Makings of a Good Meeting*, by Kevin Wolf (available at www.dcn.davis.ca.us/go/kjwolf/)

South Mountain's Vineyard Future Sketch*

Dedicated to the memory of Ed Logue, a pioneer planner who loved the Vineyard, and to Herb Abrams—it was his idea to do this!

Introduction

Martha's Vineyard is being overwhelmed by its own desirability and prosperity. Land values are rising and life is changing. Most people are acutely aware of the changes, but our political leadership has not formulated a collective vision of the future. Instead, our posture is defensive. We're like a fortress under siege, weakly trying to fend off the inevitable.

But is it inevitable? Can we imagine a future that would make it possible for our children and grandchildren to enjoy the Vineyard the way we have? Can we muster the strength and resolve to tackle the hard issues and forge a satisfying future that serves the interests of all? Perhaps . . .

During the last days of January 2000, South Mountain Company's twenty-five employees, along with several friends and planning experts, spent two days thinking about the future of the Vineyard. Our goals were several:

- to sketch the outline of a future we would like to see
- to decide what commitments we, as a company, are willing to make to achieve such a future
- to share our findings with the Vineyard community in ways that might inspire similar inquiries, create dialogue, and lead to action

These are the findings of one small group of people who live and work here. We share these thoughts in a spirit of unity and collaboration, and

---

*Presented in the form in which it was produced in March 2000.

we hope we can work together to turn this ship around. It may feel like we're mired in a sea of molasses, but perhaps that's because we haven't yet put our shoulders to the wheel . . . together.

This is only a sketch. Some of the ideas are more developed than others. Some are already in place and need only to be encouraged and/or better funded. Mostly they are offered to promote further discussion.

We gathered early on a Friday morning. After introductions, David McCullough addressed the group. He spoke about the Chagres River, which almost unhinged the building of the Panama Canal but was eventually harnessed to become a part of the engineering solution, and compared it to the river of money that is pouring into the Vineyard, unhinging a way of life. He expressed five ideas that became an important part of our discussions:

- "We need political unification."
- "We need reliable information."
- "We must redirect the river of money [that causes such harm] to restoration of community."
- "We must avoid the destination-resort mentality."
- "Our future is a design issue—it should be the result of intent rather than circumstance."

Brainstorms—What We Loved and the Forces of Change

David's remarks led us to our first activity, a brainstorming session to answer the question, What did you love about the Vineyard when you first came here (*most people*) or when you were growing up here (*a few*)? Some of those:

- The easygoing pace
- The illusion of distance—the "awayness" of the Vineyard
- The freedom and accessibility of living here—"you could always find a shack or a campsite or a caretaking gig"

- The sense of youthfulness—the feeling of the Vineyard as a "campus"
- The lack of social judgment—"jail was ok on your resume"
- The lush summer vegetation—arched canopies over Middle and North Roads
- The bond between land and people based on traditional understanding
- The egalitarian nature of the place—"the Vineyard disrupts hierarchy and mixes the common with the celebrated"
- The "island quality"—clear edges, natural boundaries
- The "grain of New England" is embodied here, a special part of the regional heritage
- The way children are valued here and the strong sense of community they are raised with
- "You used to be able to get along fine without a full time job"
- The tribal culture of the Wampanoag, the black culture of Oak Bluffs—the diversity
- "On the Vineyard it's harder to fall through the cracks"
- Four-digit calling up-island
- The discovery of new nooks and crannies each year—a small place full of new surprises and endless beauty
- The fair, Jungle Beach, the village centers, the farms and fields
- What the water all around does—the smell, the sound, the light, and the boats
- The old-timers—there were more of them
- Nobody locked anything
- "Lazy Brothers—it was ok to be lazy"
- Empty roads
- Camping at Webb's and Cranberry Acres
- The scale—"you feel bigger here"

Umm. . . feels good enough to keep, doesn't it?
After identifying the things we loved, we moved on to a sometimes (but not always) gloomier subject: the causes of change. What forces are changing the Vineyard?

Some of the suggestions:

- The power of the almighty dollar makes it hard to design in scale.
- The fishing industry has dwindled.
- Visitors and new residents seem impatient.
- New money brings new values—conspicuous consumption was once absent; now, it's accepted.
- E-mail and faxes propel the pace of change and the change of pace; technology is a "fast bridge" to the mainland.
- The landscape suffers many little losses, and they all add up.
- The Vineyard as investment—the expectation of making money from real estate.
- Traffic year-round.
- Sprawl—the disappearing edge between town and country.
- The Martha's Vineyard Commission should be a vital change agent, but it's not.
- The next generation can't live here—affordable housing is a crisis.
- "Private Property—Keep Out."
- Fewer at-home parents and fewer farmers tied to the land.
- Year-round population growth—"I don't recognize as many people anymore."
- Year-round pressure of work.
- Fewer seasonal rituals = erosion of traditional values.
- Negative unemployment—the "Help Wanted" signs suck people across the Sound.
- "Put on the brakes" thinking without complementary constructive alternatives.

Nothing too positive in that list. We did identify some positive forces of change, however, including:

- the accomplishments of the Land Bank
- the increasingly activist stance of the housing authority, the Island Affordable Housing Fund, and other housing organizations
- the Brazilians as a new cultural identity
- the value still placed on traditional economies and craft

After comparing the place we loved with the forces that were changing it, we began to consider the components of a positive future—one that might harness the forces of change to allow the Vineyard to continue to be a place we (and the generations to come) will love.

The Eight Seeds in the Apple

In the discussions and work sessions that followed, we concluded that there were eight critical seeds, each of which needed important work, and each of which might bear fruit if it were carefully nurtured. We labeled them as follows:

- achieving political unity
- creating the Vineyard Institute
- promoting adequate and appropriate housing
- preserving and enhancing rural character
- supporting the new traditional economy
- making a transportation system that works
- committing to environmental stewardship
- maintaining cultural traditions

In each of the eight areas, we characterized the problem and worked on solutions. The results of those discussions follow.

Achieving Political Unity

The Problem:
- At present, there is no effective regional political leadership. The term *regional* has a harsh ring to some, and we lack full agreement that regionalism is essential. Fear of regional unity has become a crippling illness.
- The Martha's Vineyard Commission has unusually broad powers and a strong mandate, but it is not realizing its potential for regional planning.

- County government is entirely separate from town governments and has little influence on most issues.
- Most pressures facing the Vineyard are island-wide; solutions must be the same.
- There is no "connective tissue" that joins the many effective planning, social, and cultural efforts that coexist on the Vineyard.

The Solution:

- The thirteen colonies found a way to join and still retain their individual identities; so did the diverse tribes of the Iroquois and the modern postwar European countries. Surely there is a way for the six towns on Martha's Vineyard to do so, and to learn to find regional political and planning solutions that elude us now.
- All issues should be dealt with on the most local level that is *practical and effective*, and all organizations (town governments, county government, planning agencies, et cetera) should be connected. When someone holds the strings of a large number of balloons, each has its own buoyancy and lift, but they are all held firmly. On the Vineyard, some entity needs to be holding the strings.

Therefore:

- Do an island-wide nonbinding referendum to see whether the public wishes to instruct the all-island selectmen to find a way to become a "federation," to make the individual towns a part of a greater union. We may find that the public is ahead of our political leadership on this issue.
- Strengthen the Martha's Vineyard Commission through broad institutional change and additional funding to make this organization the effective planning leader that it should be and the "connective planning tissue" for the island.
- Encourage true collaboration among year-round and seasonal residents so it becomes clear that the interests of the two groups are the same and the perceived differences are erased.

Creating the Vineyard Institute

The Problem:

- We need reliable information—physical, social, and economic—to guide future decisions.
- We are at the mercy of the "hired expert syndrome"—well-regarded experts giving polar-opposite opinions on issues depending on whom they work for.
- The many diverse and random ongoing planning efforts among the towns, nonprofits, MVC, and business community are fractured and disconnected—the left hand doesn't know what the right hand is doing.

The Solution:

- An unaligned, impartial, unimpeachable source of high-quality information and data—a think tank for Vineyard issues.

Therefore:

- Create and fund the Vineyard Institute, an organization that could develop highly accessible information to guide decision makers.
- Model the institute on the WorldWatch Institute; have an annual "State of the Vineyard" report to track progress.
- For example, on the issue of open space, the institute would work with the towns, land bank, and conservation organizations to map open-space needs (for scenic, biological, agricultural, recreational, historic, and cultural purposes). With an understanding of the real needs, open-space acquisition could be focused and targeted to best purpose.
- The institute could be expanded into an educational institution— the "College of the Vineyard"—for local public education purposes and leadership training.

Promoting Adequate and Appropriate Housing

The Problem:
- We lack diverse affordable housing options for year-round working people.
- Important long-term Vineyarders are leaving the island due to the affordable housing shortage, sapping the vitality and strength of the community and depleting the workforce.
- There is a pathetic lack of housing funding. To date we have not satisfied even so basic a need as to have an executive director for our housing authority.

The Solution:
- Vigorously pursue all avenues toward good housing options. Stepping up the efforts will give hope to the young and under-housed and empower them to imagine a future on the Vineyard and to work toward that goal.

Therefore:
- Create a compelling "case statement" that documents, in simple and graphic terms, the need and extent of the problem.
- Through the Island Affordable Housing Fund, mount a strong campaign for housing funding that results in a housing endowment (from private investment and contributions) and a permanent housing income stream from a housing bank (equivalent to the land bank) or from a rooms and rental tax and island-wide impact fees.
- Create island-wide affordable housing qualification and equity standards.
- Begin widespread initiatives and create zoning changes (like those presently happening in West Tisbury) that will create more diverse housing options, including more rental housing, more in-town apartments, more "sweat-equity opportunities," more planned communities (such as Island Cohousing), more houses moved to new sites rather than torn down, more housing "barn raisings" by tradespeople, et cetera.

• Reduce land costs through small-lot village zoning, differential tax assessments, inheritance tax adjustments with housing linkages, and resident homesites with perpetual limited-equity deed restrictions.

Preserving and Enhancing Rural Character

The Problem:
• Recent development of the Vineyard is primarily suburban rather than rural. The qualities of rural/village development patterns and suburban/sprawl development patterns are juxtaposed below:

Rural/Village	Suburban/Sprawl
productive communal land use	land as individual barrier
responsive to the seasons	separate from the natural world
modest and appropriate in scale	bigger = better
reverence for tradition	aggressive display of "newness"
celebrative of unique local qualities	homogenized; "could be anywhere"

The Solution:
• Continue and encourage the rural/village patterns of pre–World War II Martha's Vineyard.

Therefore:
• Rather than continuing sprawl, identify and create new village nodes or expand existing ones to combine housing with essential commercial services in areas where there are concentrations of people living and/or working. This will make a more livable, more walkable, less congested Vineyard. Possible locations:
  « at the Blinker
  « at the junction of old County and State Road in West Tisbury
  « at Ocean Heights
  « on Chappy
  « at the Airport Business Park

« at the Morning Glory Farm intersection

« at Chilmark Center

A good existing example is the area around Tony's Market in Oak Bluffs.

- In commercial areas, promote village qualities: walkability, appropriately scaled buildings close to the street, commercial/residential mix, places to gather and interact, activity day and night, apartments above businesses, minimal parking (in the rear), sharp edges (no sprawl), dense housing nearby, and pastoral areas and green space within and outside village clusters.
- Imagine models through proactive design visioning. For example, design some new villages, redesign the North Tisbury business district, and create plans to guide growth and change.

Supporting the New Traditional Economy

The Problem:

- To be sustainable, our economy must extend beyond tourism and building.
- Our present economy does not support enough meaningful, year-round, environmentally responsible work and commerce.
- Our traditional economy, based on fishing and agriculture, has withered away.

The Solution:

- Economic development efforts should be directed toward revitalizing the agrarian economy and supporting a lively, craft-based economy dominated by small businesses and artisans.

Therefore:

- Develop craft-based apprenticeships and educational opportunities; expose island young people to traditional island crafts, trades, and livelihoods.
- Support agricultural and aquacultural start-ups and make low-interest loans available for such ventures.

- Develop new agricultural infrastructure (such as a slaughterhouse for locally raised meat and poultry products).
- Establish a center of higher education (see "Creating the Vineyard Institute," above) with a strong local cultural focus.

Making a Transportation System That Works

The Problem:
- The Vineyard has become very congested and continues to be heavily reliant on automobiles; driving is the only way to get around.
- We do not have a multimodal integrated transportation system that includes the mainland and the entire island.
- We need a financial mechanism that will allow us to dramatically ramp up public transportation.

The Solution:
- Dust off the 1996 Donaher Report that emerged from the Martha's Vineyard Transportation Task Force. Put implementation into high gear.

Therefore:
- Institute an annual island vehicle sticker ($30 for a day, $30 for a year) that will finance public transportation improvements and reduce short-term vehicle trips to the island.
- Create public transportation that is convenient, comfortable, attractive, and highly flexible (like Bermuda!).
- Support Steamship Authority solutions that integrate and truly satisfy both island and off-island port communities (and beyond).
- Encourage delivery services and other traffic-reduction strategies.
- Encourage the development of new village centers (see "Preserving and Enhancing Rural Character," above) to shorten trips and make the island more walkable.

Committing to Environmental Stewardship

The Problem:

- We don't know what our carrying capacity is; therefore, we won't know when we've fouled our nest beyond repair.
- We don't have reliable and incontrovertible environmental information on which to base our decisions.
- We don't know what our true open-space needs are.
- We have not yet committed to the most environmentally sustainable methods for dealing with water, waste, and energy.

The Solution:

- We need to know.
- We need to commit.

Therefore:

- Support the production of a new level of environmental information and resource mapping (see "Creating the Vineyard Institute," above).
- What we do know is that we have a sole-source aquifer that must be vigorously protected; therefore, we must commit to the most environmentally advanced waste and sewage disposal methods that are financially feasible (a good example is the Nantucket composting system; that system should be implemented here).
- Support our conservation organizations fully in all their creative approaches to open-space acquisition and maintenance of biological diversity. Their dedicated efforts have improved the quality of our life and landscape.
- Promote "undevelopment"; identify developed areas important for open-space needs or rural character preservation for future undevelopment using a land bank right-of-first-refusal mechanism.
- Pioneer effective mechanisms for the transfer of development rights.
- Support the development of locally produced renewable energy.

Maintaining Cultural Traditions

The Problem:
- As the young depart, we lose generational continuity, and thus we lose our cultural traditions, our passed-down oral history and knowledge about land and water, and our stories of the past.

The Solution:
- Make every effort to preserve the knowledge of our elders and to keep our young here.

Therefore:
- Perpetuate cultural traditions and oral knowledge through demonstration, application, and mentoring.
- Enable generational continuity by supporting income diversity (see "Promoting Adequate and Appropriate Housing," above) and deferring inheritance taxes until a property leaves a family.
- Identify and cultivate youthful opinion leaders and bring them into the island decision-making process.
- Take lessons from the Wampanoags about preserving heritage and passing down oral history and knowledge.

The South Mountain Company Commitment

Having wound our way through the eight issue areas and come to some broad consensus conclusions, we shifted our focus. We asked ourselves the following question: Given the vision of the Vineyard that we see before us, what can we, as a company, do to further and support the future we've identified?

The discussion was lively. Ideas flowed easily. The result was a series of actions to pursue as a company:

For large homes, institute an internal South Mountain development-of-regional-impact (DRI) process. Apply this to houses over thirty-five hundred square feet. Establish a set of criteria and a DRI checklist modeled after the

Martha's Vineyard Commission checklist, review the checklist with clients in advance of accepting large projects, and be public about our intentions (both in terms of our internal system and in encouraging the adoption of an Martha's Vineyard Commission DRI rule for all large Vineyard houses). Accept only those large projects that have more benefits than detriments. Promote simplicity, frugality, modesty, energy efficiency, and beauty. Educate our clients about true Vineyard values.

Use our money to support our goals. All agree that we should step up our charitable contributions to 10 percent of profits. All agree that we should focus our donations on local issues—housing, cultural, agriculture, and fishing projects in particular—but it was suggested that we should still leave room for a portion to go to off-Vineyard causes. We could consider starting a foundation in conjunction with other Vineyard businesses, especially those that are not now making substantial contributions (to increase the overall contribution rather than just diverting funds from one place to another). We could extend the reach of our funds by using a portion as social venture capital to support/establish/loan money to newly formed "risk" ventures that further our Vineyard goals.

Do more work in the realm of planning and development feasibility. We might fund a housing needs assessment and case statement and contribute it to the community. We might collaborate with other design firms to do public design visions for areas like Five Corners or the North Tisbury Business District. This could extend to mapping new ideas for the Vineyard, which might include new village nodes, open-space and resource data, identification of areas for wind generation of local electricity, et cetera.

Revise the South Mountain Company statement of purpose and goals. Review and update. Make it more specific; use it to communicate our Vineyard values to potential clients.

Offer leadership to the design/build community. React to the "builders' line" about building caps threatening livelihoods. Form a positive builders' association or, even better, design/build association. Be out front and vocal about our support for a sensible future.

Offer to improve the West Tisbury Dumptique and add a building materials exchange. Take this on as a South Mountain project. Discuss with the Refuse District, the West Tisbury Board of Health, and the Dumptiquers to determine needs. Offer to assemble a plan and proposal for the Dumptique and a building materials exchange. If it is embraced (or even encouraged), do it!

Continue to plan diverse affordable housing projects. Specifically, look toward developing in-town, adaptive reuse or renovation projects and continuing with planned community work in more walkable locations.

Use the experience of older South Mountain employees as an educational tool beyond the company. Consider a variety of training options, such as apprentice programs. Build bridges of interdependence with communities that will otherwise regard us with envy when resources become scarce in the future (such as New Bedford).

Improve our use and recycling of construction and demolition waste. Explore whether it is possible to get a gypsum tub grinder for Martha's Vineyard. Work on creating a building materials exchange at the dump. Make more of our waste lumber into kindling, and systemize better. Try to sell high-quality salvage wood scraps as a resource on the Internet. Support the establishment of an island-wide trash composting facility. Replace porta-potties with portable job-site composters.

Use brush as a building material. Return to the consideration of how waste brush, tree cuttings, and clay (all plentiful on the Vineyard) can be made into building materials, such as clay/chip blocks.

Step up our commitment to reclaimed, renewable, energy-efficient materials and systems in our projects. Take this commitment farther down the path to sustainability.

Create a timber bank to replace the wood we use. Trees do not easily grow into usable timber on Martha's Vineyard. Consider investing in woodlots elsewhere in New England and growing timber using sustainable practices.

Have meetings to follow up, extend, and work on all of the above. Perhaps have short company meetings each month after board meetings. Bring in outside speakers whenever possible for inspiration and guidance. Expose the community at large to these people when appropriate.

Figure out how best to bring our findings to a larger audience. The employees of South Mountain Company are conveyers of a vision and willing participants in achieving it. Consider working with the high school students to illustrate and flesh out these ideas. Show our leaders what's possible. "If the people lead, the leaders will follow."

Toward a Future We Can Love (Not Just a Future We Can Live With)

David McCullough concluded his Friday remarks by saying, "We may not win, but at least we'll know we were there." Isn't this why we undertake such efforts—to navigate and steer rather than drift on winds of change? Whatever the outcomes, at least we'll have committed to the effort. We'll know we were there.

We learned from this endeavor. We learned how powerful it is when a large community of interest thinks together in a facilitated discussion. We learned that:

- *There's no crisis here.* We have not crossed a line and destroyed our nest. We can change what we wish to, as long as we take a long view.
- *What is happening to the Vineyard is neither new nor unique.* Our problems are the same as those facing all beautiful places.
- *Our problems can be solved.* Witness the issue of open space. It took a multifaceted effort to create the success we are having with open space, including active conservation groups, large financial contributions, the creation of a land bank, et cetera.

- *We must take a similarly creative approach to other issues.*
- *If we look closely, it's clear that we've made a good start.* Lots of good planning has already been done; it's important to remember what we've accomplished on the Vineyard as well as what needs to be done.
- *The most effective tool for change is models that make radical new approaches seem commonplace.* Our Future Sketch is one example.

At the end of the session Tom Chase of the Nature Conservancy and Armando Carbonell of the Lincoln Institute (and former founding executive director of the Cape Cod Commission), neither of whom had much previous experience with South Mountain, summarized what they had observed. Both were impressed by the breadth of thinking and the ideas that emerged from a group of people in a for-profit business. They were generous with both ideas and spirit and said two things loud and clear:

- "On the Vineyard, now is the time."
- "Get out there and be a role model—this is important stuff."

We are grateful to David McCullough, Tom Chase, Armando Carbonell, Tony Lewis, Bruce Coldham, and Sanford Evans for their assistance and participation, and we are grateful to one another for the synergistic collaboration that resulted from thinking hard together.

Buckminster Fuller once said, "You never change things by fighting the existing reality. To change something, build a new reality that makes the existing reality obsolete." We wish to offer these ideas and commitments to the island in the hope that it may help in some small way to move us all to passionately embrace an optimistic, equitable, and sustainable future for Martha's Vineyard. And to build a new reality.

NOTES

Chapter 1

1. Bernard Kamaroff, *Small-Time Operator* (Laytonville, CA: Bell Springs, 1991).
2. Mitchell is still farming, with his wife, Clarissa, on land that has been in her family for centuries. The Allen Farm Sheep and Wool Company has become one of the emblematic new breed of Vineyard farms.
3. Some of the authors who are, in my view, doing a good job of examining globalization are Jeff Gates, William Greider, Charles Handy, Paul Hawken, Marjorie Kelly, and David Korten.
4. Economist Schumpeter (1883–1950) felt that companies had to be organized for innovation and constant change, which he termed "creative destruction."
5. Marjorie Kelly, *The Divine Right of Capital—Dethroning the Corporate Aristocracy* (San Francisco: Berrett-Koehler, 2001), 110.
6. This phrase appeared in *A Voice Crying in the Wilderness: Notes from a Secret Journal,* the last book by this anarchist novelist, essayist, and environmental protector. It was published in 1989, the year Ed Abbey died.
7. I think I heard Forbes say this in a talk sponsored by the Vineyard Conservation Society.
8. Charles Handy, *The Hungry Spirit* (New York: Broadway Books, 1999), 121.
9. Lee wrote this in his edits of an early draft of a chapter of this book. We can be certain that his comments did not indicate that my writing made clear my awareness of the slightness of my knowledge.

Chapter 2

1. Daniel Boorstin, *The Americans: The National Experience* (New York: Vintage Books, 1965), 21.
2. William Greider, *The Soul of Capitalism* (New York: Simon & Schuster, 2003), 13.
3. Marjorie Kelly, *The Divine Right of Capital—Dethroning the Corporate Aristocracy* (San Francisco: Berrett-Koehler, 2001), xvi.
4. Ibid., 186.
5. Greider, *The Soul of Capitalism,* 272.
6. These quotes come from the minutes of a series of meetings held in the spring of 1986.
7. Pitegoff's help was invaluable; we were lucky to find him. His nondoctrinaire attitude was particularly reassuring. He'd never heard of such a long waiting period, for example, but he endorsed it because he understood that we were, essentially, designing a house for ourselves that we had to be comfortable living in. We could always remodel

later. The structure has stood the test of time. The five-year waiting period has turned out to be none too long.

We were fortunate, too, to have a smart and open-minded accountant, Gerry Tulis of Tulis Miller in Boston. He had never done accounting for employee-owned cooperatives, but he took it upon himself to become an expert. These days he is able to advise companies that are seeking to make the transition, and he handles the accounting for a number of worker-owned cooperatives.

8. Kelly, *The Divine Right of Capital*, 152.
9. John Logue and Jacquelyn Yates, *The Real World of Employee Ownership* (Ithaca, NY / London: ILR Press, 2001), 14.
10. Ibid., 73.
11. Peter Pitegoff, "Worker Ownership in Enron's Wake—Revisiting a Community Development Tactic," *The Journal of Small and Emerging Business Law* 8 (2004): 239–59.
12. Logue and Yates, *The Real World of Employee Ownership*, 171.
13. Pitegoff, "Worker Ownership in Enron's Wake," 255.
14. This document is available on the Internet at http://www.red-coral.net/WorkCoops.html.
15. I've lost track of where I originally read this.
16. I've lost track of where I read this, too.
17. Charles Handy, *The Elephant and the Flea: Reflections of a Reluctant Capitalist* (Boston: Harvard Business School Press, 2001), 129.
18. Jeff Gates, *Democracy at Risk: Rescuing Main Street from Wall Street* (Cambridge, MA: Perseus, 2001), lii.

Chapter 3

1. Herman E. Daly, *Beyond Growth: The Economics of Sustainable Development* (Boston: Beacon Press, 1996), 167.
2. Jamie S. Walters, *Big Vision, Small Business: The Four Keys to Finding Success & Satisfaction as a Lifestyle Entrepreneur* (San Francisco: Ivy Sea, 2001), 58.
3. George Gendron, "America's Favorite Hometown Businesses," *Inc.* (July 2002): 20.
4. Charles Handy, *The Hungry Spirit* (New York: Broadway Books, 1999), 106–7.
5. Gunter Pauli, *Upsizing* (Sheffield, UK: Greenleaf, 1998), 121.
6. Paul Hawken, *Growing a Business* (New York: Fireside, 1987), 94.
7. Handy, *The Hungry Spirit*, 107–8.
8. Gendron, "America's Favorite Hometown Businesses," 20.
9. Sarah Susanka, *The Not So Big House* (Newtown, CT: Taunton, 1998), 5.
10. It is beyond the scope of this book to consider large enterprise, but, ironically, Fritz Schumaker's classic *Small Is Beautiful* has a very thoughtful analysis of how large businesses can work well, and in 1973 Schumacher fully anticipated the problems of the global economy and recommended local control and oversight of large businesses. See chapter 19, "New Patterns of Ownership."

11. Information about the Rule of 150 was assembled from three places: *The Tipping Point: How Little Things Can Make a Big Difference* (Boston: Little, Brown, 2000), by Malcolm Gladwell; the Web site CommonSenseAdvice.com, run by a woman with the poetic name Lyric Duveyoung; and from the W. L. Gore Web site, http://www.gore.com.
12. Gladwell, *The Tipping Point*, 180.
13. William Greider, *The Soul of Capitalism* (New York: Simon & Schuster, 2003), 88.
14. Virginia Postrel, *The Substance of Style* (New York: HarperCollins, 2003), 10.
15. Dave Thomas and Michael Seid, *Franchising for Dummies* (New York: Hungry Minds, 2000), 104.
16. Tom McMakin, *Bread and Butter: What a Bunch of Bakers Taught Me about Business and Happiness* (New York: St. Martin's, 2001), 52.
17. Ibid., 63.
18. Ibid., 57.

Chapter 4

1. John Elkington, *Cannibals with Forks: The Triple Bottom Line of 21st Century Business* (Stoney Creek, CT: New Society Publishers, 1998).
2. Wayne Norman and Chris MacDonald, "Getting to the Bottom of 'Triple Bottom Line,'" *Business Ethics Quarterly* (March 2003).
3. At this point on the Vineyard, there seem to be more skunks than dogs and people combined. They have no natural predators here that I know off. Pedal to the metal is the only form of control.
4. These statements come from company minutes from February 1993.
5. Peter M. Senge popularized the term in his seminal book about organizations, *The Fifth Displine: The Art & Practice of the Learning Organization* (New York: Doubleday / Currency, 1990).
6. Paul Wachtel, *The Poverty of Affluence* (New York: Free Press, 1983), 284.
7. Alfie Kohn, *No Contest: The Case Against Competition—Why We Lose in Our Race to Win* (New York: Houghton Mifflin, 1992), 48.
8. Ibid., 49.
9. Ibid., 22.
10. Ibid., 8.
11. Ibid., jacket blurb on the inside back cover.
12. Jeffrey Hollender and Stephen Fenichell, *What Matters Most: How a Small Group of Pioneers Is Teaching Social Responsibility to Big Business* (New York: Basic Books, 2004), 264.
13. Robert Putnam, *Making Democracy Work: Civic Traditions in Modern Italy* (Princeton, NJ: Princeton University Press, 1993), 160.
14. Tom McMakin, *Bread and Butter: What a Bunch of Bakers Taught Me about Business and Happiness* (New York: St. Martin's, 2001), 86.

15. In *The Monk and the Riddle: The Art of Creating a Life While Making a Living* (Boston: Harvard Business School Press, 2001), author Randy Komisar says that his job, in business negotiations, was "to find intersections of interest . . . not differences, but commonalities." Negotiations are an opportunity to get to know people, to learn from them and about them, to learn what they care about. It's a time to build trust, not to win battles or, as Komisar says, "play poker."
16. James C. Collins and Jerry I. Porras, *Built to Last: Successful Habits of Visionary Companies* (New York: Harper Business, 1997), 228.

Chapter 5

1. Peter Wolf, *Hot Towns: The Future of the Fastest Growing Communities in America* (New Brunswick, NJ: Rutgers University Press, 1999), 1.
2. Richard Florida, *The Rise of the Creative Class: And How It's Transforming Work, Leisure, Community and Everyday Life* (New York: Basic Books, 2002), 6.
3. These three groups have remarkable stories. The oldest, of course, is the Wampanoag tribe, members of which have lived on the Vineyard for thousands of years. Their descendants still make up the majority of the population of our westernmost town, Aquinnah. Federally recognized as a tribe in 1987, the Wampanoags are now a significant economic force and continue to be an important cultural entity, even more so as they revive lost aspects of their heritage.

The African American population originated in the 1850s, when Oak Bluffs became a favorite resort area for middle- and upper-class blacks after a Methodist sect started summering here.

The Brazilians arrived more recently, but their impact has been swift and far-reaching. For a number of years they were quietly ensconced in menial hospitality industry jobs and were barely visible. As their population has grown (it is now estimated to be three to four thousand), they have become an important presence in the life of the community. They have started businesses and built churches. Their children are in the schools. My friend Charley, who is the school psychologist at one of our elementary schools, felt like he couldn't talk to the parents of many of his students, so he took a summer in Brazil to study Portuguese intensively.
4. Tom Kelley, *The Art of Innovation* (New York: Doubleday, 2001), 31.
5. The letter was published in the *Hartford Courant* in 2002.
6. Michael H. Shuman, *Going Local: Creating Self-Reliant Communities in a Global Age* (New York: Routledge, 2000), 128.
7. Charles Handy, *The Hungry Spirit* (New York: Broadway Books, 1999), 176.
8. Shuman, *Going Local*, 59.
9. Ibid., 71.
10. More detail about this area is contained in the article "Model of Economic Democracy," *Common Ground* (June 2003); see the Web site http://www.commonground.ca.
11. Thomas Frank, *What's the Matter with Kansas?* (New York: Metropolitan Books, 2004), 59.

12. Shuman, *Going Local*, 202.

13. Jeffrey Hollender and Stephen Fenichell, *What Matters Most: How a Small Group of Pioneers Is Teaching Social Responsibility to Big Business, and Why Big Business Is Listening* (New York: Basic Books, 2004), 265.

Chapter 6

1. In small business, it's much the same. According to the U.S. census of 2000, for every hundred small businesses, only two make goods like food, clothing, machinery, and toys. Roughly eight are involved in construction. For comparison, two clean offices; seven lease us our apartments or sell us our homes; four provide personal services like wedding planning, dating services, and dry cleaning; three cut our hair or give us manicures; and two amuse us at recreational sites.

2. Christopher Alexander, *The Nature of Order Book Two, The Process of Creating Value* (Berkeley, CA: The Center for Environmental Structure, 2002).

3. We also periodically survey past owners and take our medicine. What's wrong and what's right about your house? What would you do differently? What should no new house be without? What should no new house be cursed with? Our clients give us honest replies. Sometimes they tell us things we wish we didn't have to hear, but the information is critical to our future success. One couple, for whom we had built a passive-solar house that included a problematic dyed-concrete floor, answered our question about what houses shouldn't be cursed with by saying, "A floor that comes off on your socks." Soon that concrete was covered with tile.

4. Before 1850, most American houses were made with braced timber frames. But once sawmills were commonplace, mass-produced nails were available, and construction subsystems were developed that used small-dimension wood, houses became faster and cheaper to build. The artful and durable timber-framing methods fell into disuse.

   In the 1970s a few visionaries, led by Ted Benson and Ed Levin in New Hampshire, began a revival. It has taken hold. The results of timber framing are often beautiful and graceful, full of spirit and strength. The simultaneous requirements of large scale and precise detail make this work unusually attractive to craftspeople. It's also congenial; preparing timbers involves a lot of handwork that people can do gathered in close quarters, either in a shop or in the field.

5. Dr. Gro Brundtland, quoted in Jeffrey Hollender and Stephen Fenichell, *What Matters Most: How a Small Group of Pioneers Is Teaching Social Responsibility to Big Business, and Why Big Business Is Listening* (New York: Basic Books, 2004), 81.

6. Stewart Brand, *How Buildings Learn* (New York: Viking, 1994), 166.

7. Brand, *How Buildings Learn*, 167.

Chapter 7

1. This quote came from an Island Affordable Housing Fund brochure for the "Raising the Roof" campaign. For details, contact the Island Affordable Housing Fund, P.O. Box 4769, Vineyard Haven, MA 02568.
2. As of this writing, the median area income for a family of four in Dukes County is $52,900. Therefore, someone making 150 percent of that is making approximately $80,000.
3. David McCullough, at an Island Affordable Housing Fund fund-raising dinner during the summer of 2003.
4. Good sources for further information about cohousing are the Cohousing Association of the United States Web site (http://www.cohousing.org) and the following two books: Kathryn McCamant and Charles Durrett, *Cohousing—A Contemporary Approach to Housing Ourselves* (Berkeley, CA: Ten Speed Press, 1994), and Chris Hanson, *The Cohousing Handbook: Building a Place for Community* (Vancouver, BC: Hartley & Marks, 1996).
5. For more information about composting toilets see David DelPorto, *The Composting Toilet System Book* (Concord, MA: Center for Ecological Pollution Prevention, 2000).
6. For more information see http://www.mvca.com.
7. He was making the interesting combination of *dysfunctional* and *feng shui*. Feng shui is part of an ancient Chinese philosophy of nature. It is mainly concerned with understanding the relationships between nature and ourselves so that we might live in harmony within our environment, both indoors and out.
8. "Jenney Lane Opponents Are Wrong," *Vineyard Gazette*, April 2, 2004.

Chapter 8

1. For more information see http://www.capewind.org.
2. Tom Hale, letter to the editor, *Vineyard Gazette*, July 11, 2003.
3. Bill McKibben, "Serious Wind," *Orion Magazine* (July/August 2003), 15.
4. "It Will Only Hurt a Little Bit," *The Broadside* (West Tisbury, MA), #41 (January 24, 2001).
5. This quote comes from the Island Affordable Housing Fund video *Preserving a Way of Life*; for details, go to http://www.iahf.vineyard.net.

Chapter 9

1. Arie de Geus, *The Living Company: Habits for Survival in a Turbulent Business Environment* (Boston: Harvard Business School Press, 2002), 108.
2. Peter M. Senge, *The Fifth Discipline: The Art & Practice of the Learning Organization* (New York: Doubleday / Currency, 1990).

3. Marjorie Kelly, "'The Legacy Problem: Why Social Mission Gets Squeezed Out of Firms When They're Sold, and What to Do about It," *Business Ethics* 17, no. 2 (Summer 2003).
4. Ibid.
5. Leslie Christian, at the Legacy Project gathering, Minneapolis, 2002.
6. Stewart Brand, *The Clock of the Long Now: Time and Responsibility—The Ideas Behind the World's Slowest Computer* (New York: Basic Books, 1999), 2.
7. Ibid., 157.
8. Peter Pitegoff, "Worker Ownership in Enron's Wake—Revisiting a Community Development Tactic," *The Journal of Small and Emerging Business Law* 8 (2004), 255.
9. Ibid., 257.

Chapter 10

1. Tachi Kiuchi and Bill Sherman, *What We Learned in the Rainforest: Business Lessons from Nature* (San Francisco: Berrett-Koehler, 2002), viii.
2. William Greider, *The Soul of Capitalism* (New York: Simon & Schuster, 2003), 22.
3. "Can This Man Change the World?" *Guardian Weekly*, October 15–21, 2004, 17.

Appendix 2

1. Arie de Geus, *The Living Company: Habits for Survival in a Turbulent Business Environment* (Boston: Harvard Business School Press, 2002), 191.

SELECTED BIBLIOGRAPHY AND RESOURCES

Adams, Frank T., and Hansen, Gary B. *Putting Democracy to Work*. San Francisco: Berrett-Koehler, 1987.

Alexander, Christopher. *The Nature of Order*. Berkeley, CA: The Center for Environmental Structure, 2002.

Alexander, Christopher, Ishikawa, Sara, and Silverstein, Murray. *A Pattern Language*. With Max Jacobson, Ingrid Fiksdahl-King, and Shlomo Angel. New York: Oxford University Press, 1977.

Alperovitz, Gar, Williamson, Thad, and Imbroscio, David. *Making a Place for Community: Local Democracy in a Global Era*. New York: Routledge Press, 2002.

Berry, Wendell. *Another Turn of the Crank*. Washington, DC: Counterpoint, 1995.

Berry, Wendell. *A Continuous Harmony*. Washington, DC: Shoemaker & Hoard, 1970.

Berry, Wendell. *The Unsettling of America—Culture and Agriculture*. San Francisco: Sierra Club Books, 1997.

Boorstin, Daniel J. *The Americans: The Democratic Experience*. New York: Vintage Books, 1973.

Boorstin, Daniel J. *The Americans: The National Experience*. New York: Vintage Books, 1965.

Brand, Stewart. *The Clock of the Long Now: Time and Responsibility—The Ideas Behind the World's Slowest Computer*. New York: Basic Books, 1999.

Brand, Stewart. *How Buildings Learn*. New York: Viking, 1994.

Brown, Lester R. *Eco-Economy: Building an Economy for the Earth*. New York: W. W. Norton, 2001.

Calthorpe, Peter, and Fulton, William. *The Regional City—Planning for the End of Sprawl*. Washington, DC: Island Press, 2001.

Cohen, Don, and Prusak, Laurence. *In Good Company: How Social Capital Makes Organizations Work*. Boston: Harvard Business School Press, 2001.

Collins, James C., and Porras, Jerry I. *Built to Last: Successful Habits of Visionary Companies*. New York: Harper Business, 1997.

Collins, Jim. *Good to Great: Why Some Companies Make the Leap . . . and Others Don't*. New York: Harper Business, 2001.

Cramer, James P. *Design Plus Enterprise: Seeking a New Reality in Architecture*. New York: The American Institute of Architects Press, 1994.

Cronon, William. *Changes in the Land: Indians, Colonists and the Ecology of New England*. New York: Hill & Wang, 1983.

Daily, Gretchen C., and Katherine Elliston. *The New Economy of Nature: The Quest to Make Conservation Profitable*. Washington, DC: Island Press, 2002.

Daly, Herman E. *Beyond Growth: The Economics of Sustainable Development*. Boston: Beacon Press, 1996.

Daly, Herman E., and Cobb, John B., Jr. *For the Common Good*. Boston: Beacon Press, 1989.

De Geus, Arie. *The Living Company: Habits for Survival in a Turbulent Business Environment*. Boston: Harvard Business School Press, 2002.

DePree, Max. *Leadership Is an Art*. New York: Dell, 1989.

DePree, Max. *Leadership Jazz*. New York: Currency Doubleday, 1992.

Douthwaite, Richard. *The Growth Illusion*. Gabriola Island, BC: New Society, 2002.

Duany, Andres, et al. *Suburban Nation: The Rise of Sprawl and the Decline of the American Dream*. New York: North Point Press, 2000.

Elkington, John. *Cannibals with Forks*. Stoney Creek, CT: New Society, 1998.

Florida, Richard. *The Rise of the Creative Class: And How It's Transforming Work, Leisure, Community and Everyday Life*. New York: Basic Books, 2002.

Frank, Thomas. *What's the Matter with Kansas?* New York: Metropolitan Books, 2004.

Frankel, Alex. *Word Craft*. New York: Crown, 2004.

Garreau, Joel. *The Nine Nations of North America*. New York: Avon Books, 1981.

Gates, Jeff. *Democracy at Risk: Rescuing Main Street from Wall Street*. Cambridge, MA: Perseus, 2001.

Gates, Jeff. *The Ownership Solution*. Portland, OR: Perseus, 1998.

Gladwell, Malcom. *The Tipping Point: How Little Things Can Make a Big Difference*. Boston: Little, Brown, 2000.

Greider, William. *The Soul of Capitalism*. New York: Simon & Schuster, 2003.

Halstead, Ted, and Lind, Michael. *The Radical Center: The Future of American Politics*. New York: Anchor Books, 2001.

Handy, Charles. *The Elephant and the Flea: Reflections of a Reluctant Capitalist*. Boston: Harvard Business School Press, 2001.

Handy, Charles. *The Hungry Spirit*. New York: Broadway Books, 1999.

Hanson, Chris. *The Cohousing Handbook: Building a Place for Community*. Vancouver, BC: Hartley & Marks, 1996.

Hawken, Paul. *The Ecology of Commerce. A Declaration of Sustainability*. New York: HarperCollins, 1993.

Hawken, Paul. *Growing a Business*. New York: Fireside, 1987.

Hawken, Paul, Lovins, Amory, and Lovins, L. Hunter. *Natural Capitalism*. Boston: Little, Brown, 1999.

Hayden, Dolores. *Redesigning the American Dream: The Future of Housing, Work, and Family Life*. New York: W. W. Norton, 1984.

Hesselbain, Frances, et al. *The Community of the Future*. San Francisco: Jossey-Bass, 1998.

Hiss, Tony. *The Experience of Place*. New York: Knopf, 1990.

Hock, Dee. *Birth of the Chaordic Age*. San Francisco: Berrett-Koehler, 1999.

Hollender, Jeffrey, and Fenichell, Stephen. *What Matters Most: How a Small Group of Pioneers Is Teaching Social Responsibility to Big Business, and Why Big Business Is Listening*. New York: Basic Books, 2004.

International Forum on Globalization. *Alternatives to Economic Globalization: A Better World Is Possible*. San Francisco: Berrett-Koehler, 2002.

Jacobs, Jane. *The Nature of Economies*. New York: The Modern Library, 2000.

Jenkins, Robin. *The Road to Alto*. London: Pluto Press, 1979.

Kamoroff, Bernard. *Small-Time Operator*. Laytonville, CA: Bell Springs, 1991.

Kelley, Tom. *The Art of Innovation*. With Jonathan Littman. New York: Doubleday, 2001.

Kelly, Eamonn, Leyden, Peter, and members of Global Business Network. *What's Next? Exploring the New Terrain for Business*. Cambridge, MA: Perseus, 2002.

Kelly, Kevin. *New Rules for the New Economy: 10 Radical Strategies for a Connected World*. New York: Penguin Group, 1998.

Kelly, Marjorie. *The Divine Right of Capital—Dethroning the Corporate Aristocracy*. San Francisco: Berrett-Koehler, 2001.

Kiuchi, Tachi, and Sherman, Bill. *What We Learned in the Rainforest: Business Lessons from Nature*. San Francisco: Berrett-Koehler, 2002.

Klein, Naomi. *No Logo*. New York: Picador USA, 2002.

Kohn, Alfie. *No Contest: The Case Against Competition—Why We Lose in Our Race to Win*. New York: Houghton Mifflin, 1992.

Komisar, Randy. *The Monk and the Riddle: The Art of Creating a Life While Making a Living*. Boston: Harvard Business School Press, 2001.

Korten, David C. *The Post-Corporate World: Life After Capitalism*. San Francisco: Berret-Koehler, 1999.

Kunstler, James H. *The Geography of Nowhere: The Rise and Decline of America's Man-Made Landscape*. New York: Simon & Schuster, 1993.

Lakoff, George. *Don't Think of an Elephant!* White River Junction, VT: Chelsea Green Publishing, 2004.

Logue, John, and Yates, Jacquelyn. *The Real World of Employee Ownership*. Ithaca, NY / London: ILR Press, 2001.

McCamant, Kathryn, and Durrett, Charles. *Cohousing—A Contemporary Approach to Housing Ourselves*. Berkeley, CA: Ten Speed, 1994.

McKibben, Bill. *Hope, Human and Wild: True Stories of Living Lightly on the Earth*. Boston: Little, Brown, 1995.

McMakin, Tom. *Bread and Butter: What a Bunch of Bakers Taught Me about Business and Happiness*. New York: St. Martin's, 2001.

Moe, Richard, and Wilkie, Carter. *Changing Places: Rebuilding Community in the Age of Sprawl*. New York: Owl Books, 1997.

Morrell, Margot, and Capparell, Stephanie. *Shackleton's Way: Leadership Lessons from the Great Antarctic Explorer*. New York: Penguin Books, 2001.

Morrison, Roy. *We Build the Road as We Travel*. Philadelphia: New Society, 1991.

Pauli, Gunter. *Upsizing*. Sheffield, UK: Greenleaf, 1998.

Postrel, Virginia. *The Substance of Style*. New York: HarperCollins, 2003.

Putnam, Robert D. *Making Democracy Work*. Princeton, NJ: Princeton University Press, 1993.

Ray, Paul, and Anderson, Sherry Ray. *The Cultural Creatives: How 50 Million People Are Changing the World*. New York: Harmony Books, 2000.

Schumacher, E. F. *Small Is Beautiful: Economics as If People Mattered. 25 Years Later . . . with Commentary*. 25th anniversary ed. Point Roberts, WA / Vancouver, BC: Hartley & Marks, 1999.

Schwartz, Peter. *The Art of the Long View*. New York: Currency Books, 1996.

Schwartz, Peter. *Inevitable Surprises*. New York: Gotham Books, 2003.

Senge, Peter M. *The Fifth Discipline. The Art & Practice of the Learning Organization*. New York: Doubleday / Currency, 1990.

Shuman, Michael H. *Going Local: Creating Self-Reliant Communities in a Global Age*. New York: Routledge, 2000.

Susanka, Sarah. *The Not So Big House*. With Kira Obolensky. Newtown, CT: Taunton, 1998.

Thomas, Dave, and Seid, Michael. *Franchising for Dummies*. New York: Hungry Minds, 2000.

Wachtel, Paul. *The Poverty of Affluence*. New York: Free Press, 1983.

Walters, Jamie S. *Big Vision, Small Business: The Four Keys to Finding Success & Satisfaction as a Lifestyle Entrepreneur*. San Francisco: Ivy Sea, 2001.

Wolf, Peter. *Hot Towns: The Future of the Fastest Growing Communities in America*. New Brunswick, NJ: Rutgers University Press, 1999.

INDEX

Pages in italics refer to illustrations.

CHELSEA GREEN PUBLISHING

the politics and practice of sustainable living

A Complete Guide to Green Building Options

DANIEL D. CHIRAS

SUSTAINABLE LIVING has many facets. Chelsea Green's celebration of the sustainable arts has led us to publish trendsetting books about innovative building techniques, regenerative forestry, organic gardening and agriculture, solar electricity and renewable energy, local and bioregional democracy, and whole foods and Slow food.

For more information about Chelsea Green, visit our Web site at www.chelseagreen.com, where you will find more than 200 books on the politics and practice of sustainable living.

The New Ecological Home:
A Complete Guide to
Green Building Options
Daniel D. Chiras
ISBN 1-931498-16-4
$35.00

Planet

PINHOOK
Finding Wholeness in a Fragmented Land

Janisse Ray
AUTHOR OF
Ecology of a Cracker Childhood

Pinhook: Finding
Wholeness in a
Fragmented Land
Janisse Ray
ISBN 1-931498-74-1
$12.00

Food

REVISED AND EXPANDED EDITION

WHOLE FOODS
COMPANION

a guide for adventurous cooks, curious shoppers, and lovers of natural foods

DIANNE ONSTAD

Whole Foods Companion
Dianne Onstad
ISBN 1-931498-62-8 (pb)
$35.00
ISBN 1-931498-68-7 (hc)
$60.00

People

This Organic Life
Confessions of a Suburban Homesteader

Joan Dye Gussow
A CHELSEA GREEN LIVING BOOK

This Organic Life:
Confessions of a Suburban
Homesteader
Joan Dye Gussow
ISBN 1-931498-24-5
$16.95